Intel Inside New Mexico
A Case Study of Environmental and Economic Injustice

SouthWest Organizing Project
Albuquerque, New Mexico
1995

In collaboration with the
Electronics Industry Good Neighbor Campaign
(A joint project between the Southwest Network for
Environmental and Economic Justice and the
Campaign for Responsible Technology)

Electronics Industry Good Neighbor Campaign (EIGNC):
SouthWest Organizing Project, Albuquerque, New Mexico
People Organized for the Defense of Earth and her Resources
(PODER), Austin, Texas
Tonatierra Community Development Institute,
        Phoenix, Arizona
Silicon Valley Toxics Coalition, San José, California

We would like to thank the following:
Lenny Siegel of Pacific Studies Center, Mountain View, California
Sanford Lewis of the Good Neighbor Project
Greenpeace USA

We extend special appreciation to the Jessie Smith Noyes Foundation
for funding assistance and critical support of this effort.

General Assistance funding which aided in this Report provided by:
The Public Welfare Foundation
The North Shore Unitarian Universalist Veatch Program

Funding for printing costs provided by:
Chicano Studies, University of New Mexico
Presbyterian Church (USA)

Publisher's Note: *Intel Inside New Mexico* looks like it will be a "never ending story." The main body of this text was written and published in March, 1994. The "Aftermath" updates the reader to March 1, 1995. If you want to keep up, read *Voces Unidas*, the SouthWest Organizing Project's quarterly news magazine ($10 per year suggested donation, send to address below).

Table of Contents

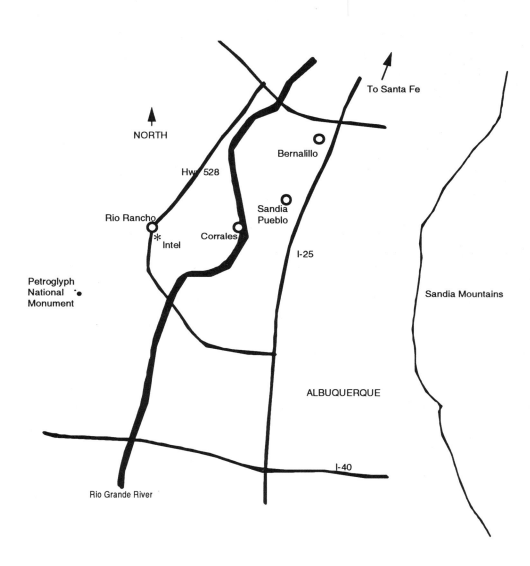

NORTH

To Santa Fe

Hwy 528

Bernalillo

Sandia
Pueblo

Rio Rancho

* Intel

Corrales

I-25

Petroglyph
National
Monument

Sandia Mountains

ALBUQUERQUE

I-40

Rio Grande River

On April 1, 1993 (April Fools Day) New Mexico media trumpeted the $1 billion expansion by Intel for the Rio Rancho, New Mexico facility. 1000 jobs were to be created. Slowly, New Mexicans (over 50% people of color) are recovering from the initial euphoria as the environmental and community impacts become known. The total expansion and retooling is now estimated at $2 billion.

Intel Chairman Gordon E. Moore had received assurances from the "Environmental" Vice-President, Albert Gore, that the permit process for the plant would be streamlined: Intel uses "some pretty noxious chemicals. . . it's important. . . we move rapidly and. . . make sure unnecessary red tape didn't stand in the way. . . We just wanted to make sure the administration understood the problem." Had the meeting gone the other way? "We would have gone outside the U.S." (*Albuquerque Tribune*, April 5, 1993)

Why Rio Rancho, New Mexico? "We're going where we get the best deal," explained Richard Perlman, head of the Intel selection team. (*San José Mercury News*, April 4, 1993)

Acronyms Used In This Book

AMD: Advanced Micro Devices
AMREP: American Real Estate and Petroleum
BAAQMD: Bay Area Air Quality Management District
CEM: Continuous Emissions Monitoring
CEO: Chief Executive Officer
CEQA: California Environment Quality Act
CERCLA: Comprehensive Environmental Response
    Compensation Liability Act (Superfund)
CFC: Chlorofluorocarbons
CPU: Central Processing Unit
DCE: Dichloroethylene
DEC: Digital Equipment Corporation
DOS: Disc Operation System
DRAM: Dynamic Random Access Memory
EPA: Environmental Protection Agency
FAB: Fabrication plant
FLP: Floating Point Unit
FTZ: Foreign Trade Zone
GATT: General Agreements on Tariffs and Trade
GE: General Electric
GTE: General Telephone and Electronics (Lenkurt)
IBM: International Business Machines
IRB: Industrial Revenue Bond
NAFTA: North American Free Trade Agreement
NMED: New Mexico Environment Department
NMITC: New Mexico Investment Tax Credit
NMMFA: New Mexico Mortgage Finance Authority
NMOHSB: New Mexico Occupational Health and
    Safety Bureau
NPOCs: Non-Precursor Organic Compounds
PC: Personal Computers
PNM: Public Service Company of New Mexico
POCs: Precursor Organic Compounds
R&D: Research and Development
ROD: Record of Decision
RRUC: Rio Rancho Utility Corporation
SARA: Superfund Amendments and Reauthorization Act
SIA: Semiconductor Industry Association
SNEEJ: Southwest Network for Environmental
    and Economic Justice
SWOP: SouthWest Organizing Project
TCE: Trichloroethylene
TVI: Technical Vocational Institute

Foreword

by Stephen Viederman, President
Jessie Smith Noyes Foundation

All of us share a dream. We want to live in a decent place where the air is clean, the water clear and drinkable, where nature, in its myriad forms, is available to us, and where our cultural heritage is preserved. We want decent jobs that pay a fair and just salary for a day's work, and that do not expose us to hazards to our health, while providing us with needed goods and services at reasonable prices. We also want to participate in the decisions that affect our lives, where we live and where we work. And we want these things not only for ourselves, but for our children and their children. In short, we want to live in healthy communities, where the quality of life is high. In the current jargon: we want to live in sustainable communities.

Sadly, that is obviously not the case for most Americans. One example will suffice. As the Citizens Fund reported in September 1992, "230 times more toxic waste was emitted in the neighborhoods near the plants [of the fifty largest industrial toxic polluters] than in the communities of the chief officers of the companies [responsible for the waste]." Seventy percent of these executives lived in communities where toxic emissions from industrial facilities were zero.

Sadly, as study after study has recently shown, the burden of living in unhealthy communities falls disproportionately on people of color and on working and low income people.

This report on New Mexico's purchase of Intel in Rio Rancho (or is it Intel's purchase of New Mexico, or both) is an important case study of contemporary smokestack chasing, to the detriment of community, people and the environment. This provision of significant tax breaks, waiving environmental regulations, dislocating people and small businesses, among other things, is a practice that even the *Wall Street Journal* is calling into question. But apparently, the word has not yet gotten around to state officials in New Mexico, and Alabama (that just bought or was bought by Mercedes Benz) and elsewhere.

This report underlines the fact that economic justice and environmental justice are one and the same thing. Economic development is not and cannot be an end in itself. Economic development must be seen and practiced as a means to the end of a healthy community, ensuring its inhabitants and their children a high degree of economic security and democracy. At the same time it must maintain the integrity of the ecological systems upon which all life and all production depends.

This report should be read and used by communities, taking action to create healthy communities; by investors, who are concerned about social responsibility; by business and industrial leaders, and the colleges, universities and professional schools that train them, who must begin to give real meaning to their recent statements about concern for 'sustainability;' and by elected representatives and government officials at local, state and national levels, who are responsible for carrying out the will of the people in a democracy. In short, there is much here for all of us concerned with a sustainable present and future.

In closing I would like to salute the courageous women and men who have been working on this issue and similar issues in communities in New Mexico and in other communities around the nation and the world. Their work is a stirring example of community democracy in action.

Executive Highlights

Money, not "good corporate citizenship" drives Intel. . . For the
people of New Mexico, the issue is survival. . . Intel CEO
Andrew Grove: "Only the paranoid survive." *page 4*

(In 1985) Intel laid off 6000 employees, abandoned the memory
business to the Japanese, and wagered its future on the 386
microprocessor. *page 7*

Chips, and Intel, are capitalism at its most creative and at its
most destructive. *page 11*

And the bottom line, money, requires the industry leader to be
even tougher and more aggressive than its competitors.
Intel has proven equal to the task. *page 11*

"Microprocessor design is a lot like Russian roulette, except you
pull the trigger, put the gun to your head and then wait two
years to find out whether or not you've blown your brains
out." And Sandoval County has signed a thirty year lease
with this company in this industry. *page 11*

Intel leads all companies in the Silicon Valley in Superfund sites
with three. *page 13*

Gordon Moore, Intel's chairman, was quoted: "It was so fast
that permits are no longer an issue.... A lot of our concerns
are over things like permitting, what with our short product
cycles. We can't afford to wait around. Getting good
cooperation from public officials is really important to us."
page 17

The DEC study found a miscarriage rate for women in the
industry at double the rate of women not exposed to the
chemicals. The IBM study disclosed a 33% miscarriage rate
for women exposed to glycol ethers. The SIA study, which
included Intel and the SIA, identified a 40% greater rate of
miscarriages for female workers exposed to glycol ethers.
page 23

"There is always the possibility that union activity could become an outgrowth of SWOP's focus on Intel, or that union activity could evolve from some other direction." (Intel Labor Relations Review Handout) *page 27*

On the South is Bernalillo County and Albuquerque, with its labor force and its sewage treatment plant. On the East is Corrales. *page 32*

With the 1980 deal, Intel, "Lord of the Chip," carved out an unincorporated fiefdom with unprecedented benefits. The 180 acre Intel Property, "owned" by Sandoval County and totally controlled by Intel, is about as unincorporated a property as you can find. *page 31*

Intel funds Synchroquartz.
Synchroquartz funds the bonds.
Sandoval County issues the bonds.
Sandoval County leases the property back to Intel for the coupon interest on the bonds.
Intel pays the coupon interest through the Trustee, Sunwest Bank, to Synchroquartz.
Synchroquartz is Intel. *page 33*

As you would expect (since Intel wrote it), the 1980 Lease was a sweetheart of a lease. The lease was for 30 years, in an industry where new generations of products and processes come every three to five years. *page 34*

"Nominal ownership of real estate" has never been more nominal than in the case of Intel and Sandoval County. *page 35*

Intel, the tenant, totally controls the site, can add improvements and raze improvements at will, can grant easements on its own, can sublease without owner approval, and can purchase the site for $1.00 at the end of the term. *page 35*

The new project, FAB 11, will be covered by a new lease, which will expire in 2023. *This represents an additional 13 years of property tax exemption and sales tax relief*, over the earlier leases, due to expire in 2010. *page 37*

Gary Parker, Intel's senior Vice President and manager of

technology and manufacturing was quoted in the February 22, 1993 *San José Mercury News*, "Pentium FAB (Rio Rancho's prize) probably will only be good for two generations (of products). Six years after it's opened, it'll look pretty old."

page 38

These bonds will be habit-forming, a bond charge card if you will, provided Intel remains a viable company.   *page 38*

Adding It All Up: Intel's Tax Breaks For The First Five Years
Industrial Revenue Bonds
    Property Tax Abatement $100 million
    Sales Tax Abatement $82 million
    NMITC $50 million
    Income Tax Abatement $15 million
    Job Training Funds $3 million
                  TOTAL: $250 MILLION

page 45

Andrew Grove, Intel CEO: "This is what we do. We eat our own children, and we do it faster and faster. We're going to move this thing as fast as we can because its the right thing to do, and that is how we keep our lead."   *page 45*

And certainly as important, the $250 million does not include the environmental and social costs of the expansion.

page 45

The report (N.M. Bureau of Business Research) pointed out that Sandoval County's 1992 tax base was $1.7 Billion; the Intel IRB deal(s) will effectively remove $2.9 Billion of improvements from the tax rolls.

page 45

At best, our public officials gave away too much. At the worst, New Mexico may face a financial and environmental nightmare.   *page 48*

For EVERY SINGLE six-inch silicon wafer processed in a FAB, the following resources are used:
        3,200 cubic feet of bulk gases
        22 cubic feet of hazardous gases
        2,275 gallons of deionized water

20 pounds of chemicals
285 kilowatt hours of electrical power
And for EVERY SINGLE six-inch wafer, the following output is produced:
    25 pounds of sodium hydroxide
    2,840 gallons of waste water
    7 pounds of miscellaneous hazardous waste
These figures are particularly shocking when we consider that Intel wants to process 5,000 **eight inch** wafers per week when the new Rio Rancho FAB is fully operational in 1995.

page 50

In November, 1993, *with FAB 11 only a construction site*, NMED found Intel in violation of emission standards, because of production at the existing FABs. *page 50*

This means, among other things, making sure that companies report hazardous waste, and storing the information. To this day, none of the "Right to Know" information in the state of New Mexico is computerized. Boxes of files crowd fire stations throughout the state, a bureaucratic nightmare for anyone looking for information—a fire hazard in the fire station. *page 51*

Corrales Comment: "Objectively, what is there about Intel's operations that Corrales residents should be concerned about. . . ?"
Intel's Richard Draper: "[Intel Environmental Safety Manager David] Shea says 'odor'." *page 55*

The well permit applications from Intel, Rio Rancho Utilities Corporation and the City of Albuquerque threatened a cumulative withdrawal of 39,500 acre feet per year from the aquifer. "(The 39,500 acre feet) corresponds to the amount of water depleted by irrigation of at least 7,600 acres and as much as 18,810 acres of farmland in the Albuquerque area. This amount of irrigated acreage may ultimately be required to be taken out of production." (State Water Engineer Eluid Martinez) *page 58*

In plain language: If Intel lowers the water table, drill deeper and pay for it yourself. *page 59*

In plain language: If Corrales is worried about water quality, put in a sewer system. Use the property tax. This, from the tax-exempt Lord of the Mesa. *page 59*

In plain language: Grow silicon chips and not green chile.
page 59

Intel is provided the infrastructure at the expense of communities such as the South Valley where people have lived for over 300 years. *page 62*

For decades, residents have lived with facilities like the treatment plant and landfills as neighbors, yet have not been afforded access to the services. For Intel the process of accommodation and access is already in full swing. *page 62*

The Pueblo officials stand firm in their conviction to protect the Rio Grande. As former Isleta Governor Verna Williamson told a roomful of Public Works officials, surrounded by Native Americans with PhDs in everything from hydrology to law: "You sent us to your schools, and now we're back."
page 65

In early 1993 Intel hired The Hirst Company, an Albuquerque Public Relations (PR) firm, to prop up its image. Later, Intel hired on Pat Delbridge Associates, Inc., a Canadian PR firm which lists major corporate polluters like Dow Chemical and Motorola among its clients. *page 67*

Intel is pushing ahead at breakneck speed, with the largest private construction project in New Mexico history. The production process to be used is "now being developed" (but it's already permitted). The impact on the air and water must still be determined. The company wants to reduce pollution at the source because it's "cheaper." The head of the Intel environmental effort considers "odor," not the toxics, the major concern. Intel Rio Rancho Manager Bill Sheppard: "The way you make advances in this business is by taking risk, by pushing the envelope, by experimenting with new methods. And it's not just in technology. It's taking risks in everything." Does anyone wonder why SWOP and the community are concerned? *page 68*

Keith L. Thompson, Intel's top executive in Oregon made the threat clear: "We can expand in a number of locations. If we don't have the incentive for Oregon, then Oregon is not competitive." *page 80*

Jonathon Krebs, New Mexico's Economic Development Department secretary responded: "I don't frankly understand present value calculations." *page 81*

The explicit assumption in the accumulation of these incentives is that 524 (52%) of the 1000 new jobs will go to people from outside New Mexico! *page 83*

"It has been my pleasure to work with Intel in the past and it will certainly be my pleasure to work with Intel in the future as well. . . .," Bill Garcia, NMEDD. *page 84*

$1.5 million for CEM is too much money; $11 million for the odor-eaters is no problem. *page 86*

"Which Intel Governor Do You Want?" ...1994 Green party candidate Roberto Mondragón. *page 87*

Timothy H. Smith, director of the Interfaith Center on Corporate Responsibility stated: "The Noyes Foundation's appearance at the Intel stockholder's meeting is the first time a foundation executive ever has taken such a step and raised tough questions with corporate management. Noyes is on the cutting edge of a new trend." *page 89*

CEO Grove: "To some people, this policy seemed arrogant and uncaring. We were motivated by the belief that replacement is simply unnecessary for most people. We still feel that way, but we are changing our policy because we want there to be no doubt that we stand behind this product." *page 91*

"'Intel on the Inside' now means that you've bought expensive, indifferent technology vended by a disdainful, deceitful manufacturer." *page 92*

Sustainable development means respect for air, water, land, people and culture. The people of New Mexico demand sustainable development. *page 92*

Introduction

New Mexico is the 48th poorest state in the U.S. People of color, primarily Chicano(a) and Native Americans, comprise over 50% of the New Mexico population. The unemployment rate of people of color is over two times as great as the Anglo unemployment rate. New Mexico has often been described as a "colony" of the United States.

The military has been the dominant employer for years. After decades of abuse of the soil, air and water, the poisoning effects of Kirtland Air Force Base, White Sands Missile Range, and Los Alamos National Laboratories are only now becoming known. The costs of cleanup will be enormous. The San José Community of Albuquerque has been identified as a Super-fund site, poisoned by the currently owned General Electric (GE) plant, a military supply plant positioned next to a tradition-al rural New Mexican community in the shadow of Albuquer-que.

Extractive industries such as uranium mining, coal min-ing, even rocks for stone-washed denim jeans have been major employers, have taken natural resources, and have left the land, air and water degraded. Coal and uranium miners, over-whelmingly people of color and indigenous people, have paid the price. Native Americans, especially, have been impacted because many of these resources have been located on their tribal lands.

So many impoverished New Mexico counties have been solicited for regional and national dumps and incinerators, we have lost count. Mescalero Apache political leaders contemplate the placement of a nuclear dump on their sacred land.

Now comes Intel, high tech industry, a modern Trojan Horse, with gleaming chrome, darkened glass and, of course, stucco exteriors, and with a chemical factory inside.

> *Now Intel, Lord of the Chip, comes to the Rio Grande Valley with a $2 billion expansion. Why are they here on Silicon Mesa?*

High tech already has a sordid history in the state. The documented poisoning of over 225 workers, primarily people of color and primarily female, by General Telephone and Electronics (GTE) Lenkurt over the 20 year history of their Albuquerque plant should alert all to the potential dangers of this industry.

Environmental Racism is "racial discrimination in environmental policy making and the enforcement of regulations and laws, the deliberate targeting of people of color communities for toxic waste facilities, the official sanctioning of the life threatening presence of poisons and pollutants in our communities, and the history of excluding people of color from the leadership of the environmental movement."[1] GTE obviously considered the demographics of the labor pool when it moved to New Mexico. So did General Electric. The historical development and the demographics of New Mexico have made many of its communities subject to a new version of economic racism.

Now Intel, Lord of the Chip, comes to the Rio Grande Valley with a $2 billion expansion. Why are they here on Silicon Mesa? What happens when their out-of-state corporate managers move into Corrales, New Mexico forcing more land out of traditional agriculture as the original Chicano(a) families are forced to sell off land and water rights to pay taxes or just because the value of the land has made continued farming a cruel joke. In Intel's calculations of the effects of the proposed

2

mining of an additional nine million gallons of water per day, is the sacredness of the river in the equation? How will their water request affect agriculture? How many families will be bought off the land?

Money, not "good corporate citizenship" drives Intel. When the potential impacts of their proposals can be so damaging to all that is sacred to the indigenous people of the mesa and valley, can their blind pursuit of the bottom line be exempted from the charges of racism? What about General Electric and the San José Community; GTE Lenkurt and its workers. . . .Intel and Corrales/Rio Rancho? Maybe its just a mix of Intel's drive for profits and of Intel's "natural arrogance." But to New Mexico's indigenous peoples, Intel is not exempt from accountability.

A study published in *New Mexico Business* in March 1971 concluded: "Spanish-Americans tend to have an emotional attachment to the land (it is a part of the family) and to value intimate personal knowledge of one's own land and a continued lineal-family land ownership. They emphasize land transfer-and-use decisions based on community welfare, along with the view that the land ownership and usage established by custom are more important than those based on legal documents. . .Anglos tend to view land as a commodity to be bought and sold if the price is right and to stress maximum monetary income from land while they hold it."[2]

A report by Jeffrey Finer in 1980 quoted Bruce King, then Governor of New Mexico, and an Intel/Rio Rancho booster: "If we let the land be ruined it's usually ruined forever. But we can't afford to forget the people who need a chance to make a living. You've got to decide how high the return has to be in terms of payroll to make it worthwhile to accept damage to the land."[3]

"Good Corporate Neighbor/Citizen" posturing aside, the bottom line for Intel is money. For the people of New Mexico, the issue is survival. Ultimately, when the affected workers and the affected communities exercise their rights to question, to inspect, and to act on the political front, benefits can accrue and the health and safety of the communities, the workers, the river, and the land will be protected. Twenty-five of the 225 workers who recently settled the suit against GTE Lenkurt are dead. Seventy-five have cancer. The critical connection between the Intel expansion and its impact on people of color must be acknowledged. Is this a paranoid overreaction?

Andrew Grove, President and CEO of Intel: "Only the paranoid survive."[4]

Semiconductor Industry and Intel[5]

The semiconductor industry was born in 1947 when researchers at Bell Labs in New Jersey demonstrated the first transistor. William Shockley, a Bell Lab researcher, founded Shockley Semiconductor in 1954 in his home town of Palo Alto, California. Shockley recruited eight promising young engineers and scientists to launch the new company. Transistors soon displaced the vacuum tube which was larger, more fragile and used much more energy in electronic devices.

In 1957, a group tabbed by Shockley as the "Traitorous Eight" and including Robert Noyce and Gordon Moore left the company and founded Fairchild Semiconductor backed by Fairchild Camera and Instruments of New Jersey. Fairchild soon became one of the world's top semiconductor firms. In the late fifties, Jean Hoerni of Fairchild invented the "planar process." The planar process was a substantial breakthrough which replaced the metal wire connectors on the transistor with metallic lines diffused onto the silicon semiconductor itself. In 1961, Robert Noyce first connected two transistors on a single piece of silicon. This was the first "integrated circuit." The "chip" was born. Fairchild flourished in the sixties with NASA and Department of Defense contracts. But Fairchild's New Jersey bosses had trouble controlling its California based subsidiary. Key managers left to form new companies or to join competitors.

In 1968, Noyce, Moore, and Andrew Grove left Fairchild to found Intel. By the early seventies, industry insiders were already calling Santa Clara County of California the "Silicon Valley."

Intel quickly became the technological leader of the semiconductor industry. In 1971, Ted Hoff of Intel created the first microprocessor. The chip now had been expanded with more and more transistors and circuits so that the brains of a computer could be squeezed onto just one single chip.

Designers and production engineers continued to place more transistors onto a single chip while reducing the cost of each element. Every two or three years, the chipsters managed to quadruple the number of transistors per chip. "Moore's Law" (named after Intel founder Gordon Moore) suggests that the average cost per circuit elements falls about 30% each year due mainly to the increase in transistors per chip.

Digital watches, video games and, by the mid-seventies, the personal computer began to change the leisure and the business world of the country. As the speed and power of microprocessors increased, the personal computer was finally developed into a marketable product with both personal and business applications with the Apple II.

International Business Machines (IBM) had resisted the trend toward the personal computer (PC), concentrating on mainframes and minicomputers for business and educational applications. But the market would prevail. In 1979, IBM entered the personal computer market and used Intel's 8088 microprocessor; Intel was indeed "inside."

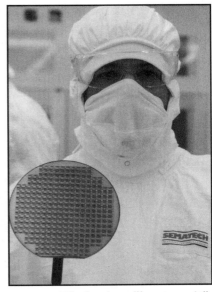

Worker in a "bunny suit" holding a silicon wafer
photo courtesy of Sematech

Intel's deal with IBM was no accident. Intel and Motorola had been the two industry leaders at the 16 bit microprocessor level. Inside Intel, the company had established "Operation

Crush," as in crush Motorola, to out-design, outsell and outmaneuver Motorola and its competing chip. Intel won with the IBM deal. Motorola has never recovered.

The IBM PC relied on "system software" developed by a young Seattle firm called Microsoft (Microsoft's Disc Operation System [DOS]). It also relied on the Intel 8088 microprocessor. The IBM PC was no technological marvel; it patched together existing technologies. But IBM was larger than all its major competitors combined and was considered the epitome of U.S. technological and managerial prowess. The IBM PC became the industry standard. "Application software" firms hurried to develop word processing systems (Wordstar), spreadsheets, and games, games, games for the IBM PC. IBM allowed other firms to build compatible computers or "clones." New firms, beginning with Texas-based Compaq, introduced a varied line of IBM clones at a range of quality, size and price. The clone-makers, competing with IBM, had to purchase the operating system from Microsoft and *the microprocessor from Intel* or firms to which Intel licensed its technology. Today there are more than 150 million PC's worldwide built around the Intel microprocessor design.

> *". . . If Intel could sell you a deliberately crippled 486 (one pictures teams of Intel engineers performing diabolical experiments on healthy 486s) for half price, why not sell a good one for the same price?"*

IBM also backed Intel with capital. At one point IBM owned twenty percent of Intel's stock. Intel retained its independence, however. In 1987, IBM sold the last of its Intel stock.

More than any other single event, IBM's anointment of Intel as the king of the microprocessors is responsible for Intel's rise to the top of the semiconductor business. Each new generation of the IBM PC and its clones has been based on a version of the Intel microprocessor family. Purchasers who upgrade to a new PC want to be able to run their old programs on the new machines requiring compatibility with the chip in the previous machine. Software firms also have a vested interest in this process.

Continued compatibility of programs such as Wordperfect and Excel with new machines keeps their products in use and marketable. Even when competitors developed higher performance microprocessors than Intel's, Intel held onto its market because most existing PCs already were powered by the Intel microprocessor design.

Even so, Intel's profits and dominance were always vulnerable. The company prospered in the early eighties but reached a crisis in 1985. In 1985 and 1986 Intel racked up cumulative operating losses of $250 million. Japanese firms were taking over the dynamic random-access memory (DRAM) business which had been based on an Intel innovation. Gordon Moore of Intel: ". . . .the world was losing its shirt in memory (DRAM). . . We tore the company down and put it back together."[6]

Intel laid off 6000 employees, abandoned the memory business to the Japanese, and wagered its future on the 386 microprocessor.

Intel won the wager. The 386 became the industry standard and a huge money machine. Intel made the decision to not license any second sources for the 386 chip and obtained a virtual monopoly on the PC market. The 386 IBM machine and its clones became the most successful product in the history of the PC industry. Application software for the machine was dynamic and provided a breakthrough in PC power and application. Intel's 1991 revenue stood at $197,000 per employee.

While Intel was gearing up 386 production, it doled out its 386 product to customers based on the past purchases. The 386 was overpriced and controlled. According to critics, Intel became:

> "egocentric, arrogant and unresponsive to its customers. . .How much Intel is hated can be judged by even its most seemingly arcane moves. . . When Intel first introduced the 486 (the successor to the 386) it included a 'floating point unit' (FLP) integrated on the 'central processing unit' (CPU). In earlier Intel microprocessors, the FLP was not included. . . .But some custom-

ers said they didn't really need an FLP so why should they pay for one? No problem, said Intel. We will sell you a 486 with the FLP disabled for 50% less. . . .Engineers, however, screamed fraud. If Intel could sell you a deliberately crippled 486 (one pictures teams of Intel engineers performing diabolical experiments on healthy 486s) for half price, why not sell a good one for the same price?"[7]

But this is Intel. The company ruthlessly exploits any advantage to generate the massive profits needed to protect those profits. Money, always the bottom line, drives the company.

Intel has not risen to the top of the high tech heap by resting on its laurels. It pursues aggressive strategies in both design and manufacturing. It propels these strategies by investing 13% of its revenues in research and development. With Intel inside the leading PC products, it can afford to. Today, the company expends these enormous resources to ensure that it will introduce new microprocessors long before other firms—with the exception of IBM, under terms that limit IBM's production—can copy them. Intel's lead time over its competitors is the monopoly basis of its money machine. Intel attempts to hang on to each monopoly product as long as possible. The design process has been accelerated. The company uses sophisticated management techniques to divide design into simultaneous activities among several groups of engineers. The company does not wait for one design to be complete before beginning the next. Intel was working on the successor chip to the Pentium even while the Pentium was in the design process. And the final production process for the Pentium, **which already has permits for air emissions,** is still in the design process!

In recent years, Intel has been in court challenging alternative designs from competitors. When Intel totally dominated the 386 market, it required purchasers of the 386 to also use Intel as the source for their 286 machines. The court battles are not over but Intel has not lost, so its dominance is not immediately threatened. But competing chipsters may be gaining, and may ultimately prevail. "In no rush to pay premium price for a

Pentium PC? If you can wait a year, chances are you'll be able to get a Pentium-compatible system powered by a Pentium-class processor from Advanced Micro Devices (AMD), Cyrix, or even Texas Instruments—and pay far less than you can imagine."[8]

Intel followed the 386 with the 486 and now the Pentium (rather than the "586," the designer chip phase...what's next, the Sextium??), which is the planned product for the New Mexico FAB (**fab**rication plant).

> *Intel followed the 386 with the 486 and now the Pentium (rather than the "586," the designer chip phase. . .what's next, the Sextium??)*

Intel has maintained the industry tradition of locating production wherever costs are lowest. Because Intel needs capacity and can afford to invest, it is developing more new FAB lines than any other firm. Communities across the U.S. have been made to actively compete for each new Intel expansion. As shown below with the New Mexico expansion, the Intel management exhibits the same aggressive and ruthless strategy in this game as with its competitors within the industry.

Intel's spatial plan follows a long established pattern. Management, research and development (R&D), and design are concentrated in the Silicon Valley which still has the highest concentration of high tech talent in the world. Important engineering teams are also located in Oregon and Arizona. As community resistance to the New Mexico expansion grew, Intel spokespersons floated tantalizing stories that R&D functions may be planned for the Rio Rancho, New Mexico facility, too. In December, 1993, the company announced tentative plans to add a $35 million R&D pilot program to the Silicon Mesa property: "Clearly, I think this (the R&D proposal) is good news for Rio Rancho, New Mexico and Intel... if we're able to go forward with it," Richard Draper of Intel.[9] Draper is a former local TV news anchor person.

Wafer fabrication,[10] the heart of chip manufacturing, takes place in Rio Rancho, New Mexico, Aloha, Oregon, Chandler, Arizona, Folsom and Santa Clara, California, and Jerusalem,

Israel. A new Intel FAB opened in Spring of 1994 in Leixlip, Ireland, to circumvent European protectionism. In each of these areas Intel has sought out the most favorable tax incentives and streamlined permitting processes. Design, testing, research and systems manufacturing take place throughout the world at the above mentioned facilities as well as in Utah, Puerto Rico, Great Britain, Germany, Hong Kong, and Japan. Final chip assembly and soldering—some of the most potentially dangerous high tech work—is done in Malaysia and the Philippines by low paid women workers.

> *"Microprocessor design is a lot like Russian roulette, except you pull the trigger, put the gun to your head and then wait two years to find out whether or not you've blown your brains out."*

Gordon E. Moore, Intel's CEO, in a newspaper OpEd piece wrote: "Intel is the world's largest manufacturer of integrated circuits, commonly known as computer chips or semiconductors. Intel and other American manufacturers of computers, semiconductors and electronic equipment battle for share in international markets that are brutally competitive, rife with innovation and constantly in upheaval. New products, considered 'revolutionary' just a few years ago, are now obsolete."[11]

Capitalism has widely touted the chip industry as a major success story, with good reason. Innovation with new products and new processes is often cited as the key component which makes the market economy hum. All reports confirm, however, that there has been an erosion of living standards and an actual reduction in leisure time in the U.S. since the 1970's for the middle and lower classes. The gap between the poor and the rich has been extended.

Notwithstanding all this, the chip industry has made a revolution. Personal computer prices have dropped precipitously as their power increased. State of the art PCs have become affordable, although only just prior to becoming outdated. This report, 15 years ago, would have required resources unavailable to the organization. And you can be certain that during the writing, revision, and printing of this report, an Intel 386 Microprocessor has been humming along making the report possible.

Chips, and Intel, are capitalism at its most creative and at its most destructive. When your major product is becoming obsolete as it goes into production, your competitive edge is always vulnerable. And the bottom line, money, requires the industry leader to be even tougher and more aggressive than its competitors.

Intel has proven equal to the task. Success to date, before the expansion: "The Rio Rancho facility is Intel's largest manufacturing operation, **producing approximately 50% of the corporation's revenue and 70% of its profits**," (Intel memo, August 10, 1993). On October 12, 1993 Intel announced a 143% jump in third quarter profits. Profits were $584 million for the quarter ended September 25, 1993. Investors reacted negatively, and the value of the stock dropped! Investors familiar with the competitive nature of this industry are aware that Intel's reign as "Lord of the Chip" is tenuous. But Intel's earnings for the fourth quarter of 1993 increased to $594 million. Earnings for all of 1993 were $2.3 billion on revenues of $8.8 billion. Intel can definitely afford to pay its share of New Mexico taxes.

Michael Slater, industry guru, in a 1992 *Upside* interview stated the problem: "In terms of market share, probably 1990 or 1991 was Intel's peak. . . .The degree of dominance and the degree of control that they have are going to continue to fade. . . ." And later in the same interview he quotes a favorite analogy: "Microprocessor design is a lot like Russian roulette, except you pull the trigger, put the gun to your head and then wait two years to find out whether or not you've blown your brains out."[12] And Sandoval County has signed a thirty year lease with this company in this industry.

For additional perspective, a word about Intel's microprocessors and their development is useful. In many ways, Intel's current dominance of the microprocessor is more a testament to Intel's manufacturing prowess than its technological strength. The development from the 286 chip to the 386, 486, now the Pentium (586) and beyond is really a quantitative movement.

Each successive chip has dramatically increased transistors, which translates into more power. Intel's manufacturing muscle is the key, then, to its bottom line—profits. Intel's David House, "Architecture is only 20 to 30 percent...the real changes occur in the silicon technology."[13] As its huge profits attest, the New Mexico facility has proven itself an efficient and reliable chip producer. Efficient production and the compatibility requirement for software users and firms is the key to Intel's dominance.

Intel's Environmental Record in Silicon Valley[14]

> "An expert is someone who has made all the mistakes there are to make in a very narrow field"..... Niels Bohr, nuclear pioneer.

In the environmental arena, Intel qualifies as an expert according to Niels Bohr's definition. Their most costly mistakes were in improper and sloppy solvent handling which has caused extensive groundwater contamination at three of their main sites in Silicon Valley. These actions prompted the U.S. Environmental Protection Agency (EPA) to list these three sites on the National Priority List (Superfund) as among the most highly contaminated in the country. In addition, Intel has historically used the Silicon Valley air basin as its private sewer by spewing several hundred thousand pounds of air pollutants into the air over the past six years.

Intel has learned some expensive lessons from these experiences. From the profit line perspective, pollution prevention is now considered a good business practice. Intel's New Mexico spokesperson Richard Draper acknowledged, "It's expensive to clean up those things that pollute and it's cheaper to cut it off at the source."[15] But a more ominous trend is also developing with the company's well orchestrated attacks on the environmental permitting process in several states, including the New Mexico expansion. Intel's increasing rush to make newer and faster chips has taken the company into the political arena. The company has convinced political officials in California and New Mexico to short-circuit long-standing environmental protections in order to build new or expanded production facilities. As always, their mantra has been: *If you don't bend your rules to accommodate our needs, we will simply go somewhere where they will.* In many arenas, this type of strong arm behavior is called *"jobmail."*

Releases for Intel's international facilities are not available since the other countries do not require toxic release reporting.

Intel's Superfund Sites in Silicon Valley

Intel leads all companies in the Silicon Valley in Superfund sites with three. Two of these sites are in Santa Clara and the third in Mountain View. Chlorinated solvents have also been discovered at a fourth Intel site in Santa Clara but the contamination has not reached Superfund status. The most serious site, if we can rank Superfund sites, is at 365 East Middlefield Road in Mountain View. The underground plume at this site is the second largest in Santa Clara County and threatens San Francisco Bay. The EPA has estimated that cleanup may take up to 60 years. Intel has also been identified by EPA as a polluter at the Hassayampa Landfill Superfund site in Maricopa County, Arizona. Intel and others were ordered by EPA in April, 1993 to begin immediate cleanup of this site.

Graphic by Laura Maclay.

Intel 365 East Middlefield Road in Mountain View

This site was the first Intel plant in the country and has

proved to be its most costly mistake. A major plume of solvent contamination was discovered at this site on September 21, 1981 containing trichloroethylene (TCE), xylene and 1,2 dichloroethylene (DCE) at levels that exceeded health standards by up to 4000 times. Three nearby wells were contaminated by this spill which was caused by leaking underground storage tanks. The plume has now spread to more than 6000 feet in length and 500 feet in depth. It has commingled with similar plumes from adjacent Fairchild and Raytheon facilities, and has merged with massive contamination from Moffett Field Naval Air Station located just across Highway 101. The so-called joint regional plume continues to spread, and now threatens to pollute San Francisco Bay. As of October, 1991 (latest available data), less than 40% of the plume was contained. EPA has estimated that it could take up to 60 years to fully clean up the site!!

The EPA first identified this site as a Superfund site in 1985. Cleanup oversight, which was being handled by the California Regional Water Quality Control Board, was shifted to the EPA in 1985 due to squabbling between Intel and the other companies. A final cleanup plan Record of Decision (ROD) was issued by EPA in 1989. This Intel site as well as the Fairchild, Raytheon and Moffett Field sites are now being managed as a regional issue by EPA.

The plume continues to spread, and now threatens to pollute San Francisco Bay. As of October, 1991, less than 40% of the plume was contained. EPA has estimated that it could take up to 60 years to fully clean up the site!!

Intel Facility 2880 Northwestern Parkway, Santa Clara

The Intel Santa Clara III facility performs chemical quality control and testing of semiconductors.

An underground plume of contamination was discovered at this site in July, 1982. It contains TCE and Chlorofluorocarbons (CFC) 113 resulting from Intel's mishandling of solvents including accidental spills and overflows. TCE was detected at almost 100 times the allowable drinking water standard in a plume that has spread 350 feet and extends 30 feet deep.

In October, 1984, the site was proposed for listing as a Superfund site by EPA. A groundwater cleanup system was installed in 1985, and a final cleanup plan was approved in 1990. Intel was ordered to extract and treat polluted groundwater until it meets drinking water standards. This is estimated to take 11 years and will cost about $600,000.

Intel Magnetics/Micro Storage 2986 Oakmead Village Court, Santa Clara

This site was occupied by Intel from 1978 to 1987 for magnetic bubble production and testing operations. Prior to that time, the site was occupied by Micro Storage which is a co-responsible party.

An underground plume of contamination was discovered in 1982, containing TCE, Trichloroethane, 1,1 DCE and CFC 113. The source is believed to be a leaking underground storage tank. TCE was detected at a concentration that exceeds drinking water standards by more than 100 times. The plume stretches 850 feet and reaches 30 feet deep.

The site was placed on the Superfund list by EPA in May, 1986. A clean up plan was ordered in August, 1991 which requires that Intel pump and treat the contaminated groundwater until it meets drinking water standards.[16]

Intel's Toxic Releases in California

In 1986, Congress passed an amendment to the Superfund law (CERCLA) that required companies to report to the public their total toxic releases to the environment. Companies were required to report discharges to air, land and water, starting with their data for 1987. There are many gaps in these laws, also known as the Superfund Amendments and Reauthorization Act (SARA Title III).

The list of chemicals to be reported to the federal government is very short, and companies only have to provide information on these chemicals when they use more than a certain amount. Intel, therefore is not required to provide information on most of the 90 plus chemicals that it uses.

The total reported releases in California for the six year period from 1987 to 1993 were 354,262 pounds or roughly 177 tons. The company's air emissions included xylene, acetone, hydrogen fluoride and sulfuric acid among others. Intel was also required to report air emissions to the Bay Area Air Quality Management District starting in 1990, with somewhat broader reporting requirements. Intel reported several additional discharges of concern including 12,200 pounds of cellosolve acetate (glycol ether, a potent reproductive toxin) and 12,200 pounds of CFCs (Freon 113-ozone destroyers) in 1990. By 1991, these reported releases fell to 3740 pounds of cellosolve acetate and 2530 pounds of CFCs.[17]

Intel was recently granted a permit by the New Mexico Environment Department for its Albuquerque facilities to dump up to 356 tons of toxic emissions into the air per year. When we compare these limits to the ten tons Intel currently reports in California, it is clear that Intel received another major environmental subsidy. Air pollution became just one less concern. (see page 52)

Pollution Prevention

Intel received one of the early wake up calls in Silicon Valley with the discovery of its groundwater contamination and its huge cleanup costs. Since that time, the Intel corporate environmental affairs department has promoted pollution prevention as a more cost effective method of doing business. With Intel's vast resources, and a corporate commitment to research and development, the company has made some strides in changing its process technology to reduce its use and waste discharge of several potent environmental and occupational hazards. Intel has been forced to pioneer the use of liquid arsenic, an alternative to the more dangerous toxic gas, arsine. They have also replaced CFCs with nontoxic alternative solvents, and have made commitments to replace glycol ethers, potent reproductive toxins. They continue to use many other toxic chemicals, however, and in some cases appear to be using more solvents as the size of the microcircuits continues to shrink!!

Permit Streamlining in California: Blueprint for the Future

On October 6 and 7, 1992, the *San José Mercury*, the *Wall Street Journal*, and many other major media outlets reported that Intel decided to build a new $400 million plant in Silicon Valley that would

> *Gordon Moore: "It was so fast that permits are no longer an issue.... A lot of our concerns are over things like permitting, what with our short product cycles. We can't afford to wait around. Getting good cooperation from public officials is really important to us."*

create 250 new jobs. To emphasize the importance of this, California Republican Governor Pete Wilson attended the ceremonies announcing the event. Company and state officials credited "streamlined regulatory permit processes" as the key to Intel's decision. Approval for new permits took only three months rather than 18 months.

Gordon Moore, Intel's chairman, was quoted: "It was so fast that permits are no longer an issue... A lot of our concerns are over things like permitting, what with our short product cycles. We can't afford to wait around. Getting good cooperation from public officials is really important to us."[18] Moore also wrote an Op-Ed article for the *San Francisco Chronicle* which claimed that the new expansion was only possible because environmental protection rules had been relaxed. Moore credited the Bush administration with persuading the EPA to relax rules that would have required two years to get its permits, or 30% longer than the 18 month life span Intel projects for its future microprocessors.

What was not disclosed in the news articles were the serious allegations that Intel had exerted pressure on the Bay Area Air Quality Management District (BAAQMD) to override the decisions of front line staff and speed the process along. In a July 22, 1992 memo, California staff engineer M. K. Lee discussed her review of Intel's application for air pollution discharges. Lee determined that Intel's activity would pose enough potential environmental risk to make the permit subject to the provisions of the California Environmental Quality Act (CEQA). This decision would have required Intel to prepare an Environmental Impact Report, a potentially time consuming process, that would require Intel to justify its proposed air pollution and demonstrate technically feasible mitigation. On July 31, 1992, a memo from Mark Kragen (a BAAQMD planning official) concurred with the staff engineer that "... it is apparent that the Engineering Evaluation for this permit application will not follow the specific procedures, fixed standards, and objective measurements...governing semiconductor manufacturing...Therefore, this permit application is not ministerial and is potentially subject to the requirements of the California Environmental Quality Act."

> *Intel was recently granted a permit by the New Mexico Environment Department for its Albuquerque facilities to dump up to 356 tons of toxic emissions into the air per year.*

These staff conclusions were reversed during the permit process after review by high-level BAAQMD officials who acquiesced to Intel's plea for "maximum flexibility." Intel was then exempted from the CEQA requirements by BAAQMD. This was after the intervention of California Governor Wilson's administration which created a "red team" to help "break through" the regulatory "red tape."

A careful review of the Intel permit file reveals that there were many reasons why the BAAQMD should have upheld the initial recommendations and required Intel to follow the requirements of CEQA. Many aspects of the permit process were discretionary and therefore subject to CEQA. Although Best Available Control Technology was triggered, no abatement equipment was recommended. Insufficient information was provided to determine whether the proposed projects would cause a significant air quality impact. Cumulative impacts from facility-wide emissions were not assessed.

In fact, the BAAQMD and local residents had reason to be concerned about the proposed air pollution permits that Intel sought. Intel's proposal included plans for 31 solvent stations, 53 wet chemical stations, 16 chemical vapor deposition chambers, 46 diffusion furnaces, and 48 positive photoresist applicators and developers.

Altogether, Intel sought permission to discharge 27.5 tons per year (55,000 pounds) of Precursor Organic Compounds (POCs) and an additional 3.4 tons per year (6800 pounds) of Non-Precursor Organics (NPOCs). These levels would be allowed even after Intel installed in-plant incineration (thermal oxidation) treatment units as well as more traditional air scrubbers.

Intel was issued it's permits from the BAAQMD, without any modification or without the Environmental Impact Report.

Currently, Intel "reports" slightly under 20 tons of air emissions in the Silicon Valley. Even when the new plant becomes operational, the total reportable emissions in Silicon Valley will likely remain under 50 tons annually. This amount pales when compared to the 356 tons permitted in New Mexico. And Intel certainly learned that it can exploit its economic muscle in the political arena with the California experience. As shown below, the site search process which Intel implemented with the Pentium FAB Sweepstakes was extraordinary.

The "Clean" Industry

As the semiconductor industry developed, the early images were of gleaming buildings and clean rooms—no smokestacks, no backbreaking labor, good working conditions. The Silicon Valley boomed and was a centerpiece of California's growth. The myth of the "clean industry" began to be exported to all corners of the world.

Beginning in the early seventies, New Mexicans began to see these simple buildings with no windows, nice lawns and trees, and no smoke stacks appearing in their neighborhoods. An economic shift was taking place in the U.S., from "dirty" industry such as auto and steel manufacturing, to "clean" industry, such as microelectronics. Microelectronics companies were the new wave of production. Inside the FABs were "Clean Rooms" where integrated circuit boards and microchips were produced. Microelectronics was a new industry and these companies needed many employees. Women from all over came to apply for these good jobs, and of course, they got them.

In New Mexico, one of the first companies to appear was General Telephone and Electric (GTE) Lenkurt. This company's starting pay for entry level workers in the early '70s was $1.75 an hour for a twelve hour shift. This was good pay for Chicanas and Native American women. Many of these women were heads of household, and they were supporting large families.

After GTE, many other electronics firms entered the state. Companies such as Motorola, Signetics (now Philips Semiconductor), Honeywell, Digital, and, in 1980, Intel. They all hired women of color to work on the production lines. These women were given little or no training, and there was not much open discussion on the chemicals they would be using for the manufacturing of the integrated circuit boards or the semiconductor chips.

Occupational health concerns first surfaced in the Silicon Valley in the mid-seventies when many woman began to suffer "mystery" diseases. At first, the women were dismissed as hysterical. But, as the evidence grew, the problems had to be taken seriously. A physician, Dr. Joseph LaDou (he is now the Chief of Occupational and Environmental Health at University of California San Francisco Medical School), began to compile illness data and published articles showing that the industrial illness rate for semiconductor workers was three times higher than the national average for all manufacturing. A workers' advocacy group—Injured Workers United—was formed in San José, California as a project of the Santa Clara Center for

Occupational Safety and Health to organize electronics workers suffering from chemical exposure. They succeeded in focusing attention on workplace hazards and reducing on-the-job exposures.

Chemical Use and Health Hazards

Health hazards associated with semiconductor process chemicals include the following:

Solvents

A number of solvents are known to cause problems of toxicity to humans. Some of the solvents used at Intel are: acetone, n-butyl acetate, ethylene glycol, 2-ethoxyethyl acetate, n-methyl-2-pyrrolidone, tetrafluoromethane, and xylene. Typical effects on workers are damage to the central nervous system, systemic intoxication, and reproductive toxicity.

Gases

Arsine, a gaseous form of arsenic, is the most toxic of the gases used in the electronics industry. Exposure to minute amounts can cause rapid death. Arsenic can be used in liquid or solid form, which also poses severe health hazards because it attacks the red blood cells.

Intel, Rio Rancho, New Mexico

Other highly toxic gases include phosphine, diborane, and silane.

Acids

The use of many acids such as hydrofluoric, hydrochloric, and sulfuric as etchants all pose significant risk of acid burns to workers that handle them in the fabrication of semiconductor chips.

The processes and chemicals described above can all cause significant health effects for semiconductor workers. Workers are usually the first to be exposed to toxic chemicals, and they are usually exposed to much higher doses than the general public. The semiconductor industry uses huge amounts of toxic chemicals, so it is not surprising that the illness rate is far higher than for many other industries.

Turning silicon slivers into integrated circuits on tiny chips is a precise process: A speck of dust on a chip is referred to as a 'killer boulder' on a highway since it can block the electronic current along the tiny circuitry. Toxic chemicals and gases are the basis of the cleanliness. But humans must still mix their labor with raw materials to produce the product. And much of the workforce is women of color. The profile fits the needs of the industry. The products become obsolete every two or three years. The female is considered a more "temporary" worker than the "male, head of household." Human hands create the chips and assemble the cellular phones. Often, the hands are those of women of color of child-bearing age.

> *The DEC study found a miscarriage rate for women in the industry at double the rate of women not exposed to the chemicals. The IBM study disclosed a 33% miscarriage rate for women exposed to glycol ethers. The SIA study, which included Intel and the SIA, identified a 40% greater rate of miscarriages for female workers exposed to glycol ethers.*

Women Workers and Births

Three separate studies have clearly identified a relationship between miscarriages and workers in the industry. Digital

Equipment Corporation (DEC) commissioned the University of Massachusetts School of Public Health to do a study published in 1986, IBM contracted with Johns Hopkins University for a study reported in 1992, and the Semiconductor Industry Association (SIA) contracted with University of California Davis which published its results in 1992. The rate of miscarriages by women employed in the semiconductor industry was compared to a control group. The results of the studies are dismal and unanimous. The DEC study found a miscarriage rate for women in the industry at double the rate of women not exposed to the chemicals. The IBM study disclosed a 33% miscarriage rate for women exposed to glycol ethers. The SIA study, which included Intel and the SIA, identified a 40% greater rate of miscarriages for female workers exposed to glycol ethers.

New Mexico Workers

New Mexico has experienced the GTE Lenkurt tragedy. Over 20 years, hundreds of workers, primarily women of color, were poisoned due to exposure to the chemicals employed along with the workers. The plant is now closed and Albuquerque officials are investigating soil contamination. Two hundred twenty-five former workers have been involved in a series of lawsuits against the company, the chemical manufacturers, and the insurer over a nine year period. As part of the final settlement, the workers received an average of $20,767. As of 1993, 25 of the 225 former workers are dead, 75 have cancer and 75 are totally disabled, according to the attorney representing them. Additional GTE Lenkurt workers, not included in the lawsuit(s), continue to contact the attorney.[19]

The GTE story is only the most dramatic example of worker poisoning. A panel on high tech was included in the "Interfaith Hearings on Toxic Poisoning in Communities of Color" held on April 3, 1993 in Albuquerque and organized by the SouthWest Organizing Project (SWOP). Employees from GTE Lenkurt, Honeywell, Sandia and Los Alamos National Laboratories and Motorola were represented:

"The testimony of the workers and
former workers of high tech industries left

the audience visibly distraught. . . Dorothy Morris spoke of the lack of ventilation in her work area despite the heated toxic chemicals she constantly worked with. The story of her anguish, pain, and suffering caused by exposure to the toxic substances was retold with only minor variations by the other members of the 'high tech' panel. . . .Eva Mueller, a worker at Sandia National Laboratories, described the list of illnesses suffered by her and her co-workers: 'Health effects seen in our small group of seven or eight individuals is astounding considering that 90% of these individuals had no prior history of health problems. . . Health effects to date are: carpal tunnel syndrome, nerve problems, reflex sympathetic distropy, headaches, fatigue, memory loss, attention switching or poor attention spans, slowed reflexes, encephalopathy (degenerative brain disease), positive MRIs (actual dead spots on the brain), hypothyroidism, adrenal gland failure, color pulps, lupus, cancer, menstrual problems, cervical precancerous tissue, reactive airway disease, multiple chemical sensitivities, high liver readings, sinus surgery, irritability, depression, anxiety, higher rates of infection.'"[20]

The SouthWest Organizing Project has surveyed over one hundred workers from the Albuquerque electronics industry. The survey was an oral questionnaire that was given one on one. The questions included length of employment, starting pay, ending pay, specific job done, chemicals used, chemicals worked around, ventilation, benefits provided by the company, and safety precautions taken by the company. The women and men surveyed were between the ages of 28-65. They were all people of color. The lengths of employment in microelectronics was between six months to 15 years.

The health survey showed that, across the board, none of the people surveyed had full knowledge of the hazards of working with chemicals. Each person's work place or station had very poor ventilation. The chemicals used in these shops were usually not labeled. People could only identify the chemicals by color-coded labels or by smell. The chemicals usually had to be manually transferred from large drums to smaller containers.

Former hi-tech workers testify about their toxic poisoning at SWOP Interfaith Hearings, April 3, 1993. *photo by Richard Khanlian*

Many spills occurred and the chemicals were left on the floors or on counter tops to evaporate into the air.

Solvent Induced Encephalopathy

Dr. Ruth Lilis has reported on solvent induced encephalopathy.[21] She suggests that there is a correlation between induced encephalopathy and solvent exposure. This particular disorder has been described as "irreversible organic mental syndrome" or "chronic organic brain dysfunction." According to the investigations done by Dr. Lilis, the symptoms that are frequently associated with this disorder are:

>
> Basic change in personality
> Basic changes affecting intellect
> Emotional imbalance
> Fatigue
> Loss of initiative
> Depression
> Emotional liability
> Difficulty in concentration
> Sleep disturbances

Intel is not GTE Lenkurt. This is now, 10-20 years later.

Some chemicals proven to be harmful to workers are no longer used. Do we know the impacts of new chemicals being used? Do the Intel workers of New Mexico know what all is being used in the production process, know what their hands are touching and their noses are breathing? Does Intel know?

The abandoned GTE/Lenkurt hi-tech plant in Albuquerque.

Microelectronics: The Temp Industry

Unionization is an abhorrence to Intel and the rest of the microelectronics industry. The rapidly changing nature of the industry makes hiring and firings the norm; the typical industry worker is a temp. Any union or worker organization will make management actions more difficult. A worker organization will key in on health and safety issues.

Thus, high tech manufacturing is almost entirely non-union. The microelectronics industry work force is said to be difficult to organize. The labor force is divided by race, nationality, and language. Intel's FABs, for example, are spread all over the western U.S. High tech workers are vulnerable to antiunion campaigns, usually based on the threat of job loss. These workers deal with the threat of plant relocation, automation, the downgrading of positions, "Quality Circles" which promote spying and turning in your fellow employees if any unionizing is mentioned, and lack of promotions.

Prof. Karen Hossfeld, of San Francisco State University, has found in her research that the semiconductor industry likes to hire small, female workers of color. This industry "assumes that women make more passive and obedient workers...have less of a vested interest in organizing... and are easier to manage."[22] The microelectronics companies think that firing large numbers of women does not impact communities (and thus generate resistance) as does a massive firing of male workers. The jobs are supposed to be short term, and it is customary for the industry to hire and fire en mass.

The threat of relocation and plant closing is not an idle one. There have been many plant closings in the United States in the last ten years. Remember that Intel closed plants and laid off 6000 workers in 1986. Many microelectronics workers have lost their jobs. There is a trend to move to lesser developed countries with new fabrication and assembly plants. In these lesser developed countries, the companies can pay lower wages and pay less attention to the environmental effects on workers and the earth. This may become more of a reality with the North American Free Trade Agreement (NAFTA) and the General Agreements on Tariffs and Trade (GATT). NAFTA will make it easier for electronics plants to relocate to communities with fewer resources for enforcement of environmental and occupational health and safety laws.

The anti-union propaganda used by Intel is a matter of fact. The following excerpts were taken from a confidential handout entitled: *Labor Relation Review Handout*– Intel Corporation, Rio Rancho, N.M., August 9, 1993, and outline the company line:

> ### From Intel's own *"Labor Relations Review Handout"*
> (dated 8/9/93 and marked *"Confidential"*)
>
> "There is always a possibility that union activity could become an outgrowth of SWOP's focus on Intel, or that union activity could evolve from some other direction.
>
> Intel's response to this potential:
>
> - "Union Prevention: operate our company in a way that our employees don't feel the need for a union.

- **Preparation in case of a union campaign**: all management players need to be trained on what to do if an organizing drive does occur.

- Train all managers/supervisors on this material.

Why Intel wants to remain nonunion:

- Retain flexibility to act quickly and decisively in making business decisions without having to negotiate with a union before making changes.

- Avoid cumbersome third party intervention (i.e.: a union) in dealing with employee discipline, grievances or other employee relations issues.

- Avoid risk of strikes that would interfere with production schedules."

The handbook also outlines what management can and cannot do in the prevention of unionization. The following are some tips from Intel's handbook for management when confronted with possible union organizing:

- "Urge employees to report union coercion.

- Remind employees of Intel's benefits, their worker policies.

- If offered a leaflet, accept it and pass it on immediately to the site manager."

For Intel, production is everything. Microelectronics plants manufacture components 24 hours a day, seven days a week. With the short life span of their products, any interference will have serious financial repercussions.

Intel uses some "pretty noxious chemicals," Intel Chairman Gordon Moore.[23] This combination of a dynamic industry with short term workers and new combinations of chemicals is especially troubling. Much of the chemical poisoning which has been identified with the industry takes place over time. Workers don't often fall over dead from an incident. Rather, the effects are cumulative over time. Often the damage may be done before the company or the worker becomes aware of the danger.

Background—Intel, Inside Rio Rancho, New Mexico

The original deal made with Intel in 1980 to put a plant in Rio Rancho, New Mexico probably made the current expansion inevitable. High tech industry had discovered New Mexico in the 1970's. New Mexico had Sandia National Labs, Los Alamos National Laboratories, a model labor pool, above average unemployment, a weak labor movement, no urban congestion to speak of, lots of resources, and some compliant politicians. New Mexico was perfect. The average worker in the GTE plant of the 1970's was a female Chicana/Latina with an 11th grade education. GTE employees started at the minimum wage.

AMREP was the original developer of Rio Rancho. At the end of the 1970's, the company was recovering from 12 years of investigations and legal battles due to misleading advertisements, mainly in New York papers. When Rio Rancho was first promoted by AMREP, the target buyer was the retiree from the

Intel expansion, FAB 11 Rio Rancho, NM

East. The "painted lawns" and misleading claims resulted in lawsuits and, ultimately, criminal proceedings.

In 1980, AMREP was looking to trade land for economic development. Bruce King was Governor of the State. Mr. King's ties to Rio Rancho go back to AMREP's entry into New Mexico. Over one-third of the AMREP original holdings directly or indirectly were purchased from the King family. The family made an estimated $7 million on the deal.[24] Mr. King had even appeared in a promotional film for the company:

> "Hi. I'm Bruce King, Speaker of the New Mexico House of Representatives, State Representative of Santa Fe County. And it certainly is a pleasure to appear here this morning on this beautiful spot and view the Sandia Mountains to the East . . . I'm sure that in the next two decades, we will expand much more and the city of Albuquerque will overflow and come right in the path of where we are standing this morning. . . Certainly we have a golden future in our wonderful land of enchantment in New Mexico and we would like to welcome people throughout our fine country to come and join us here in Albuquerque."[25]

In 1980, Rio Rancho, Sandoval County, Governor Bruce King, and Intel announced the location of a new plant just outside Rio Rancho. The new plant was subsidized with Industrial Revenue Bonds (IRBs), and other tax incentives. AMREP, at this time, owned the water system and could provide assurances of a dependable and cheap water supply. The labor force was there. The Intel plant would exemplify the "new" Rio Rancho. Now, Intel promised jobs for workers. AMREP could begin to market its development to working-class, first-time homebuyers.

Sandia Mountains east of Rio Rancho, NM

With the economic incentives, the deal for Intel was ideal. The location was up on a mesa overlooking the Rio Grande River. Directly across the river to the east, the Sandia Mountains rose majestically. Just below the plant lay Corrales, a rural New Mexican Chicano community facing an accelerated process of gentrification. Sandoval County cooperated with IRBs, effectively sheltering Intel from Property Taxes and the dreaded New Mexico Sales Tax. New Mexico had not yet discovered the dark side of high tech with the GTE experience, as the GTE employees were just beginning to understand the source of their medical ills.

The Intel Property

With the 1980 deal, Intel, "Lord of the Chip," carved out an unincorporated fiefdom with unprecedented benefits. The land consists of 180 acres and is located just south and east of the municipality of Rio Rancho. The property abuts N.M. Highway 528, the main strip of Rio Rancho.

On the south and east side of the property is a separate 30 acre parcel which has just been acquired for the planned expansion. The property was acquired from AMREP by "Intel Leasing Corporation" which may just be an Intel affiliate. This strip has been annexed by the City of Rio Rancho which will entitle the site to planning, zoning, and emergency response services. Within the 30 acres there is a joint powers agreement with Rio Rancho and Sandoval County. Both the county and the city will share responsibility for services provided to the 30 acres. Intel plans to drill one of its proposed wells, construct a one million gallon storage tank for water and construct an electrical substation on the site. A zoning change is currently being processed for the 30 acres to allow for the proposed Intel uses.

The 180 acre Intel property, "owned" by Sandoval County and totally controlled by Intel, is about as unincorporated a property as you can find. In 1985 Intel created a "Master Plan" for its property. Sandoval County officials have access to this "Master Plan" for development and land use; the public does not. This, despite the clear effects that Intel has and will continue to have on its neighbors. The lease(s) between Sandoval County give Intel, the tenant, virtual total control over all activities. The tract has now also been established as a "Foreign Trade Zone" (FTZ), which will provide more tax abatements for the company. An FTZ is, according to the National Association of Foreign Trade Zones, "an isolated, enclosed and policed area operated as a public utility, in or adjacent to a port of entry, furnished with facilities for lading, unlading. . .manipulating, manufacturing and exhibiting goods and for reshipping them by land, water or air. . . Merchandise. . .may be brought into a zone without being subject to the customs laws of the United States. . . "

On the South is Bernalillo County and Albuquerque, with its labor force and its sewage treatment plant. Intel has drawn up plans for a third lane from the South for Highway 528 to allow for access for employees from the south and for shipping and receiving. A new entrance to the plant will be funded by Intel.

On the East is Corrales. This traditional community is downhill and downwind from the plant and draws its water from private wells. And the community is changing to become an exorbitantly expensive place to live. The incoming managers and engineers for the Intel FAB are choosing the community as their preferred residence. And the proposed water right purchases will only drive more of the traditional families off the land and out of the community.

Situated on top of the Silicon Mesa, Intel's 180 acre fiefdom is politically free of entanglements which normal companies must face. At the same time, all the benefits of the infrastructure, which the company needs and which the surrounding communities can offer, are theirs.[26]

As we shall see below, Intel does not pay their share of the costs because of its special treatment under the Industrial Revenue Bonds and the leases. Intel has created an island of profits subject to almost no controls from the community and from the political entities who have the responsibility of controls. Truly, this is quite an extraordinary achievement by the company. Intel, the Lord of the Chip, has carved out a virtual fiefdom, with exemptions from regulatory controls which impact other companies. The company's reticence from entering into any "Good Neighbor Agreements" is understandable, given its position of power.

Industrial Revenue Bonds (IRBs)

The Industrial Revenue Bond (IRB) is a paper transaction whereby cash-rich Intel has Sandoval County issue IRBs which are purchased by Synchroquartz, Inc., an Intel subsidiary.

> *Intel funds Synchroquartz. Synchroquartz funds the bonds. Sandoval County issues the bonds. Sandoval County leases the property back to Intel for the coupon interest on the bonds. Intel pays the coupon interest through the Trustee, Sunwest Bank, to Synchroquartz. Synchroquartz is Intel.*

Intel funds Synchroquartz.
Synchroquartz funds the bonds.
Sandoval County issues the bonds.
Sandoval County leases the property back to Intel for the coupon interest on the bonds.
Intel pays the coupon interest through the Trustee, Sunwest Bank, to Synchroquartz.
Synchroquartz is Intel.

Because Sandoval County is the "nominal owner of the real estate" (Thomas Hughes, the attorney representing Sandoval County at the August 16, 1993 meeting of the Sandoval County Commission), the property is not subject to property taxes.

Because Intel, as the tenant, is authorized to contract on behalf of the County in the development of the "project," the cost is exempted from the New Mexico Sales Tax (Gross Receipts Tax). The 1980 project is dwarfed by the current $2 billion expansion, and by the future.

1980 Project

The 1980 project was a $40 million deal. The entire set of

documents—Lease, Resolution, the Bonds, the Indenture of Trust were written by Intel and their counsel. Sandoval County Manager Debbie Hays advised that the 1993 Bond issue is the first case where the County has had legal representation. In earlier issues Intel and Intel's Counsel prepared all the documents.

> *Given the lease provisions, Sandoval County was not even involved in the early negotiations with Intel for the current Rio Rancho expansion.*

The 1980 Lease, Bond Resolution and Indenture of Trust and the subsequent amendments were, simply put, a mess. Documents filed with the County Clerk were often unsigned. It is our understanding that the County is going to Intel to obtain signed documents, if available. The 1980 Lease covered the entire 180 acres. But when, in August, 1993, we first discussed this issue with Debbie Hays, Sandoval County Manager, she was operating under the belief that the current leases only covered the 100 acres which were already developed. And this at a time when the County was negotiating with Intel for new Bonds. Were all previous leases and Bond issues ever officially signed? If not, is the previous tax abatement legal? We may never know. The original leases have been amended and cover only the first phase already in place.

As you would expect (since Intel wrote it), the 1980 lease was a sweetheart of a lease. The lease was for 30 years in an industry where new generations of products and processes come every three to five years.

The lease committed the County to issue more bonds. Section 4.2, Article IV of the 1980 lease reads in part: "If the lessee (Intel) is not in default hereunder, the County will, on request of the lessee (Intel) from time to time, use its best efforts to issue the amount of Additional Bonds specified by the Lessee (Intel, remember). . . Any such expansion shall become a part of the Project and shall be included under this Lease to the same extent as it originally included hereunder." And this was locked in. The lease includes language whereby Intel can sue the County to meet the agreements of the lease, including, of course, the agreement to issue additional bonds.

The same language is incorporated in the present 1993 lease(s).

By 1991, the IRBs had increased to $833 million.

"Nominal ownership of real estate" has never been more nominal than in the case of Intel and Sandoval County. All the leases, past, and present, contain the same basic provisions. Intel, the tenant, totally controls the site, can add improvements and raze improvements at will, can grant easements on its own, can sublease without owner approval, and **can purchase the site for $1.00 at the end of the term**.

Even had New Mexico, Rio Rancho, and Sandoval County offered no further subsidies than the subsidies there for the taking in the 1980 Lease and its successors, Intel would have found the incentives in the lease, still going and still going, until the year 2010.

Jobmail—The FAB 11 Economic Plum

Given the lease provisions, Sandoval County was not even involved in the early negotiations with Intel for the current Rio Rancho expansion. The County was already committed to additional bonds in amounts requested by Intel, their nominal tenant. In the County's defense, given the 1980 lease, their position was weak to begin with.

But, notwithstanding the lock on the IRBs and the tax relief incorporated therein, Intel followed an unprecedented game plan in extracting additional concessions for their FAB 11 (the new Intel facility).

Make no mistake about it. "Good corporate neighbor" posturing aside, Intel is a rough, tough corporation in a brutal industry. Despite its current dominance, competitors are only one new chip, one new process or agreement, away from taking over.

For Intel the issue was clear: "We're going to build where Intel gets the best deal," said Robert Perlman, Intel senior president in charge of the site selection team for the computer giant.[27]

Intel played the field with the plant development. In May 1992, the Intel site selection team visited six sites, all adjacent to current plants. The states under consideration were California, Oregon, Arizona, New Mexico (Rio Rancho), Utah, and Texas. The site selection team let their local employees make their pitch for each location, and the company told each state to make their best offer. Early in 1993, Intel issued a "secret report," and circulated it to each of the states. New Mexico and Rio Rancho, with its King connection, were pressured even more to enhance their package.

Community vs. community; worker vs. worker; it's all rather cynical. Negotiations with companies to save jobs or for expansions are no longer unusual but are the norm. Communities are desperate to retain and obtain jobs after the destructive go-go years of the 1980's. Politicians and workers in the affected communities are effectively played off against one another, as concession after concession is extracted.

The unseemly rush to provide tax breaks, loosen environmental safeguards, and provide other subsidies is a measure of how desperate communities in this country now are to obtain jobs at any price. In the FAB 11 case, Intel used an extraordinary ploy—to lay everybody's cards on the table, as it were, and then to push for one last best and final offer to extract maybe a few more bucks in tax abatements. Given the success with FAB 11, this process is certain to be used again. The calculating, cynical process, used in the 1992 California expansion and perfected in the FAB 11 sweepstakes, appears to set a new low in "corporate jobmail."[28]

The New Mexico package, when announced, was estimated at $114 million or $114,000 per job. This figure, however, totally underestimates the true cost in lost revenues, given both the increase in the bond issue to $2 billion from $1 billion and the tax relief when looked at in its totality. In addition to tax incentives and job training funds, the New Mexico program built on benefits to employers which were vested by the changes in the Workers Compensation Law in New Mexico a couple of years ago. Even Public Service Company of New Mexico (PNM) gave the company a rate break. The Paseo Road was built, a massive public works project, dislocating many people. More roads are promised and planned. Water and water rate promises have been made.

Sandoval County did not totally rubber-stamp the additional bond issue. The "clean high tech industry" was better understood by this time. The community of Corrales had smelled the problems of this neighbor for too long. Community groups like SWOP and regional organizations like the Southwest Network for Environmental and Economic Justice (SNEEJ) were better organized and had their own resources to help communities deal with a backyard Lord of the Silicon Mesa. And, in the County's defense, the previous leases already committed the County to additional IRBs. In the end the decision was made to break out the property involved into two leases; one to cover the 1993 Bond and improvements, and one to essentially supersede the previous leases.

Intel has come out, as usual, a winner. The new project, FAB 11, will be covered by a new lease, which will expire in 2023. **This represents an additional 13 years of property tax exemption and sales tax relief,** over the earlier leases due to expire in 2010. The County has publicly stated that the new lease, which will

> *"We're going to build where Intel gets the best deal,"* Robert *Perlman, Intel senior president.*

cover the improvements in place prior to the expansion, will still expire in the year 2010.

With great fanfare, Sandoval County has trumpeted the two major concessions which Intel grudgingly accepted ("I wouldn't call it highway robbery. But we were surprised by the demands." Barbara Brazil, Intel Spokesperson):

- The County has the right to make inspections of the project. (The landlord can inspect the property!!)

- Intel agrees to pay supplemental rent to the County in the amount of $5.5 million over a period of 17 years.

The community may be able to utilize the inspection right, when the next "incident" occurs. The funds are a small payoff, given the tax breaks. Intel states that the cost of a plant is quadrupling with every new development. The public has to question the wisdom of the 30 year bonds covering a facility which has a life expectancy of three to five years. A life which may be extended by another infusion of funds and then for perhaps five more years. When FAB 11 is out of date, Intel can pull out or come back to the County with—figure it out—another bond proposal. These bonds will be habit-forming, a bond charge card if you will, provided Intel remains a viable company.

Gary Parker, Intel's senior Vice President and manager of technology and manufacturing was quoted as saying, "FAB 11 (Rio Rancho's prize) probably will only be good for two generations (of products). Six years after it's opened, it'll look pretty old."[29]

A view of the Intel plant from Corrales, N.M.

Remember that this business "is a lot like Russian Roulette...except that you wait two years to find out whether or not you've blown your brains out." With a 30 year lease, Intel's neighbors can expect to watch nervously while the company pulls the trigger some five to six times. Assuming it's a six-shooter, the odds are just plain scary.

Intel's Future On Silicon Mesa

And, Intel can fail. New generations of products are developed every two to three years. In the worst case scenario, the company could wind up in bankruptcy court and the County with the property. Since the County has no control over the property, there is no protection regarding what you get. At best, there will be a property with improvements which can be razed and then resold. A Superfund site? It's not unprecedented in the Intel history. Today, Albuquerque owns the former GTE site. The site contamination has not been fully determined.

And possibility of contamination does exist. Intel's "environmental spokesperson" at the August 16, 1993 Sandoval County Commission meeting talked about the flawed technology of the 70's and the 80's to account for the Intel Superfund sites in Silicon Valley. That was then, he said, now we know better. But it's not just nine million gallons of water a day that's proposed for the expansion—new chemicals, old chemicals, new combinations of chemical processes. With each new generation of products and processes—coming every two to five years—in this industry, new chemicals, new combinations. . . And then the industry spokespersons come forward and cite the lessons they have learned. Perhaps in the year 2010, the spokesperson will be speaking to a Native American Council, explaining that the Superfund site in the Silicon Mesa was due to the flawed technology of the 90's.

The company uses 90 different chemicals in various combinations. Intel is driven by profits in a create and destroy industry, with no vision but the bottom line.

And a major chemical "incident" can create a Superfund site overnight.

> "We don't see it as risky. We think we'll be getting a 30-year partner. Intel has proven it can stay on the leading edge, which would give us some security." (Statement by Jonathon Krebs of the N.M. Economic Development Department)[30]

> "IBM REPORTS QUARTERLY LOSS OF $8 BIL-
> LION, Big Blue Slashes Dividend, Plans 35,000
> more Job Cuts" Headline, *Albuquerque Journal*,
> Wednesday Morning, July 28, 1993.

In the meantime, some of the 1000 jobs will trickle down and 350 tons of emissions trickle down over the Rio Grande Valley and Corrales. And the County's position is that "we can't make more stringent regulations than the feds."

The County is supposed to be the owner ("nominal") of the property. Can an owner not require a tenant to meet some reasonable and substantive requirements regarding the use and inspection of the leased property?

The Cost of the IRBs

Property Tax

We assume that the $2 billion expansion would be put on the property tax roles, if Intel "owned" the plant. The estimated property tax abatement is $20 million per year or $100 million over five years. A Bureau of Business and Economic Research of the University of New Mexico Report estimates the total property tax abatement of the $2.9 billion in the IRBs issued at $30 million per year.[31]

Sales Tax

New Mexico Gross Receipts Taxes are levied at 5.5% on improvements in Sandoval County (remember Intel is not within the city limits of Rio Rancho where the levy would be higher). 5.5% of $1.5 billion equals $82.5 million. ($1.5 billion is used because the actual construction of the buildings is subject to the gross receipts tax even under the IRBs. But Intel will use the New Mexico Investment Tax Credit to avoid this tax, too.)[32]

New Mexico Investment Tax Credit

With the IRBs a given, one of the problems facing New Mexico State officials was to find more taxes which could be forgiven or credited.

The New Mexico Investment Tax Credit (NMITC) is structured to give the corporate beneficiary relief from the sales tax. A company "Invests in New Mexico" and creates jobs. 5% of the investment can be used to offset New Mexico taxes.

But for Intel, the IRB structure and financing already spelled relief. So the NMITC was amended by the State Legislature and the King administration to cover jobs related to manufacturing machinery investment regardless of the IRB financing and to provide an innovative and unique method for IRB users to capture the tax forgiveness. With the amendment, *Intel could use the credit to offset payment of withholding taxes collected from its employees and due to the state.* Of all the tax subsidies, this one seems the most insidious (and costly). Was this an "Intel Amendment" (or Kingtel Amendment)?

> *With the amendment, Intel could use the credit to offset payment of withholding taxes collected from its employees and due to the state. Of all the tax subsidies, this one seems the most insidious (and costly).*

Intel workers owe and pay New Mexico income taxes. The employer, Intel, withholds the estimated tax from the employee's paycheck. So far, this is just like any other employer.

But, instead of forwarding the tax to the state, Intel appropriates the funds as a New Mexico Income Tax Credit. The employees of the company appear to pay their fair share of New Mexico income taxes. But the taxes never get to the state. The general New Mexico taxpayer foots the bill. The estimate on the cash cost to New Mexico taxpayers (from the N.M. Tax & Revenue Department) is $50 million on the new expansion (1000 jobs X $50,000 per job).

In the case of the withholding funds, the funds are not even a tax obligation of the company which was certainly the intent of the original law. These are the income tax obligations of the employees! For New Mexicans (and especially for Intel employees) this is particularly insidious and deceitful. You get a job; you pay taxes. . . that's part of it. Working class people complain about taxes but know that their contributions fund schools, roads, transfers to the very poorest of their neighbors. But for cash-rich Intel to grab their employee's income taxes, owed the state, is a mockery of basic justice in taxation.

Technical Vocational Institute, Rio Rancho/Intel Campus. See footnote 26, page 95.

And Intel officials must be laughing about the NMITC as a development policy. New Mexico doesn't even get the employee taxes under this economic development tool.

Corporate Income Tax Relief

Another recent change in the tax code was designed to provide a tax reduction on the corporate income tax for companies who are manufacturers with the majority of their sales out of state (Intel, Kingtel??). Intel fits this profile nicely, with almost all sales out of state. The N.M. Tax and Revenue Department estimates the first year savings for Intel of approximately $3 million for the company. (Assuming that Intel prospers, the tax loss will increase.) Conservative to a fault, and keeping to the five year term, tax reduction benefits are estimated at $15 million.[33]

Job Training Funds

Intel has been and is expected to continue to be a large user of job training funds. Under this program, the state agrees to pay 50% of the starting wages of new employees for a period of up to six months during the training of the new employee. An award of $1 million in such funds was made to Intel in May, 1993. The plan for the Intel expansion goes into 1996. You can expect Intel to continue to utilize these training funds (this is their second $1 million award of funds). Estimate $3 million over the five year period.

Other Incentives in the New Mexico Deal

Foreign Trade Zone

Intel was also granted "Foreign Trade Zone (FTZ)" status in July of 1993. The trade zones are designed (surprise, surprise) to allow companies to avoid taxes and import duties. The City of Rio Rancho was granted five such zones. Rio Rancho awarded the contract to operate the FTZ to the New Mexico Foreign Trade Zone Corporation, a subsidiary of AMREP (second surprise this paragraph) in August, 1993. Lynn Yaple, an employee of AMREP and of the N.M. FTZ Corporation was quoted: "They're doing it to save money for their bottom line, but it's also job retention. It's an economic development tool and will be marketed. . .just one more thing we can offer, along with lots of sunshine, a good labor force and a good quality of life."[34] According to Ms. Yaple in October 1993, Intel had not yet applied for the exemptions. For Intel to turn its back on a tax break would be slightly out of character. The company will participate. Future tax abatement will be substantial.

Worker's Compensation

New Mexico's revamping of the Worker's Compensation program was another important plus for the state. The reform has been trumpeted as a labor-corporate reform, and New Mexico organized labor was involved in the changes. But the bottom line, even for labor, was the usual threat of jobs leaving, or as in the Intel jobmail, not showing up. Chuck Reynolds of the N.M. District Council of Carpenters, AFL-CIO, a labor member of the reform team stated: "Worker's comp doesn't do a worker any good if he doesn't have a job. The cost of workers' comp was such that our employers were losing jobs to other states." Changes were initiated in 1985, with four revisions culminating in the 1990 law. The basic idea was laudable, especially from the intuitive public attitude that you need to take the lawyers and the doctors out of the loop. The aim was to reduce business insurance costs and provide some minimal protections for the workers. The changes appear to be successful from the corporate standpoint.[35]

Insurance costs are coming down. But protections for workers are definitely eroded as well as benefits reduced. Attorney participation has been reduced but at some cost to many workers. The reforms are based on the establishment of an administrative procedure and Workers' Comp judges who administrate the claims and cases. The injured worker has the assistance of an ombudsman in the claimant process and there is a mediation process. But the ombudsman is not allowed to represent the worker at the mediation process, where "the worker is facing an insurance adjuster who knows the system backwards and forward. The average claimant has a high school education and earns $5.00 to $6.00 an hour," attorney Donald Vigil.[36] Attorney fees are limited to $12,500 in the administrative process, assuming that the mediation process is unsuccessful in resolving the issue. Jordan Fox, a journeyman carpenter for 27 years, isn't impressed. He injured his back on the job in September, 1991: "All I know is I'm living on pain pills to survive, I've got $7000 in unpaid medical bills, and I'm losing my house. I'm going to end up living in the back of my truck."[37]

Adding It All Up[38]

Intel builds a plant with IRBs, hires a new employee, mixes the new employee's labor with foreign goods, gets 50% of the employee's wages from New Mexico taxpayers, collects the New Mexico income tax due and keeps it. New Mexico literally ran away with the FAB 11 sweepstakes. Thomas Hughes, Sandoval County's counsel in the IRB negotiations, told SWOP, "They (State officials) gave away the ranch."

Total cost to the taxpayers of the state is $250 million in direct tax abatement and subsidies in the *first five years alone*.[39] This translates to $250,000 per job based on the 1000 jobs to be created. And the jobs produce a product which will be obsolete after the five years!! (Or earlier if we accept the company's premise.)

On January 27, 1994, Intel announced that the successor chip to the Pentium, code-named P6 (AKA the Sextium—see above), will be on the market in 1995, the very year the FAB 11 will go into production. Andrew Grove, Intel CEO (Chief Executive Officer): "This is what we do. We eat our own children, and we do it faster and faster. We're going to move this thing as fast as we can because its the right thing to do, and that is how we keep our lead."[40]

The $250 million does not include the FTZ tax savings. The $250 million does not include the rate reductions from PNM, contract support tax breaks, the other utility savings with the Rio Rancho Utility Corporation (RRUC) and the Albuquerque Sewage Treatment Plant.

And certainly as important, the $250 million does not include the environmental and social costs of the expansion.

Furthermore, some of the tax breaks which were designed to lure Intel will be available to other firms. The University of New Mexico's Bureau of Business and Economic Research raised the same concerns in a report issued to the Legislative Finance Committee of the N.M. House of Representatives in September 1993.[41] The report pointed out that

Adding It All Up Intel's Tax Breaks For The First Five Years

Industrial Revenue Bonds	
Property Tax Abatement	$100 million
Sales Tax Abatement	$82 million
NMITC	$50 million
Income Tax Abatement	$15 million
Job Training Funds	$3 million
TOTAL	**$250 MILLION**

Sandoval County's 1992 tax base was $1.7 billion; the Intel IRB deal(s) will effectively remove $2.9 billion of improvements from the tax rolls. Brian McDonald, Director of the Bureau was quoted, "I guess what we're mainly critical of is that no one has a thorough analysis of the benefits and costs of the 'Intel deal.' And it may be that the benefits outweigh the costs. . .We grant industrial revenue bonds with property tax forgiveness to a business. . .and we don't look at the additional costs on the state for roads and water and sewage projects."[42]

The report, *New Mexico Tax Study*, Phase II certainly reinforces the fact that New Mexico gave away far too much in the FAB 11 deal.[43] The study shows that, before all the tax abatements documented above, **the corporate tax burden in New Mexico is lower than other southwestern states including California.** And the $250 million in lost revenues are only a fraction of the cost: "...increased economic activity may produce an increase in population, which in turn produces a need for more public schools and school teachers, a need for new or expanded landfill, sewer and water treatment facilities, new roads and increased maintenance costs associated with roads, more police services, more fire protection services, more restaurant inspections, increased park maintenance cost, expanded youth programs, and increased demand for classes at local universities."[44] And need we add: "Models frequently assume that the environmental consequences of development will be negligible. It is assumed that current environmental standards are sufficient, and that development will not result in future cleanup costs. This assumption may not be accurate. (New Mexico) has Superfund sites associated with development once assumed safe."[45]

Wendy's will benefit from the Intel expansion.

Benefits

The semiconductor industry has one of the lowest job multipliers of any industry: 1.9. The multiplier means that for the 1000 jobs, the community can expect 900 to be spun off.

There will be jobs created. Intel refuses to provide any information on their employees so we can only speculate how many of the jobs will go to New Mexicans. Many of the highest paying managerial and technical jobs can be expected to go to men and women from out of the state.[46]

The worker's taxes to the state, as shown above, will mostly go to Intel through the NMITC.

Sumitomo is a job spin-off. Sumitomo Sitix Silicon Inc., is a Japanese owned high tech company and is a subsidiary of the giant Sumitomo conglomerate. It is recognized as a leader in the production of silicon wafers used in the semiconductor industry. The company has announced it will locate a new plant in Albuquerque to supply silicon wafers to Intel. But the City of Albuquerque is packaging a deal with IRBs similar to Intel's! Sumitomo has received a $67 million Industrial Revenue Bond from the City of Albuquerque. They are expected to have a payroll of $5.3 million and they are supposed to provide the city with 170 jobs—manufacturing jobs. If the job spin-offs are structured with similar tax abatements, the job multiplier is a farce. With the IRB financing, property tax abatement, sales tax abatement, NMITC abatement, all will come with it. And Sumitomo is almost as thirsty as Intel, planning to use 1.5 million gallons of water per day when fully operational!

AMREP stands to benefit a great deal as Rio Rancho grows. The company is already the largest homebuilder in the state. In 1991, AMREP constructed 511 homes compared to 139 by the #2 homebuilder.[47] The New Mexico Mortgage Finance Authority (MFA) was created in 1975 and is a quasi-government agency. The MFA generates funds by selling "tax-free" bonds and passing on the interest rate savings to first-time homebuyers. In the period 1982 to 1991, Sandoval County (and this is almost entirely Rio Rancho and AMREP) generated 26.2% of the MFA loan purchases. **This is seven times greater than the number which the population would have predicted.** Governor King has appointed David King, his nephew, as the current chairman of the MFA. AMREP has successfully changed its image and is making huge profits. Intel can properly take some of the credit for the growth. But the costs of the development must be paid by someone. On December 28, 1993, N.M. State Senator Joe

Carraro sent a "Stop the insanity" FAX to local governments on Albuquerque's Westside, including Rio Rancho. He called for a homebuilding moratorium. Carraro was quoted by *The Observer*: "When we get the money for schools in Rio Rancho, it will be two or three years before they are built. What do we do in the meantime. . .? It's going to take two hours to drive from Rio Rancho to work, and we'll have dual sessions in school. We can't ask our children to learn in that environment."[48]

And Intel pays no taxes to Sandoval County.

Governor King, many legislators, and other public officials will also benefit from the impression of the growth of a clean industry in the state. There may be political contributions as well. These are short run gains, well-suited to the political arena.

At best, our public officials gave away too much. At the worst, New Mexico may face a financial and environmental nightmare.

Intel and the Environment—
The Rio Rancho Expansion

Because of the need to keep in front of the competition, another key step for Intel was Senator Jeff Bingaman's intervention and a critical meeting with Vice-President Gore. Dr. Craig Barrett of Intel met with Gore, the point person on environment for the new Democratic administration, in March of 1993. Senator Bingaman's assistance in arranging the meeting could only be a plus for New Mexico. Intel, always pushing for an edge, wanted and obtained assurances from Gore that the permit process would be streamlined and not hampered by "cumbersome" environmental regulations. Intel was playing the same game nationally that it was pursuing in New Mexico. Here the issue was the location in the U.S. or overseas. The assurance was obtained. Intel Chairman Moore was quoted after the big announcement: Intel uses "some pretty noxious chemicals...it's important...we move rapidly and...make sure unnecessary red tape didn't stand in the way. (The Clean Air Act) is several hundred pages . . . The life cycle of some of our products is 18 months. . . We just wanted to make sure the administration understood the problem. They were very supportive of this kind of industry." Had the meeting gone the other way? "We would have gone outside the U.S."[49] Economic Jobmail was never more explicit. To reiterate: Senator Bingaman's assistance certainly helped the New Mexico cause.

INTEL KIT FOR CORRALEÑOS

EAR PROTECTOR FOR TURBINE ENGINES' HARMONIC HUM.

NOSE GUARD FOR SOLVENT CLOUD.

EYE PATCH TO BLOCK IMPOSING MONOLITH.

ADDITIONAL WATER TO SUPPLEMENT WATER TABLE.

WATER

Cartoon by Kent Blair, reprinted from Corrales Comment.

The environmental permits were streamlined to allow Intel to begin production of Pentium as soon as possible. There was no process to allow for public input into the environmental decisions surrounding Intel's latest expansion. Intel received assurances from the New Mexico Environment Department (NMED) in early 1992 that it would receive permits for the proposal; this was still the jobmail phase of the expansion. "NMED supports the concept of locating Intel's proposed FAB 11 at the Rio Rancho campus. Further, NMED's Air Quality Bureau is committed to working with Intel to ensure the timeliness of the permitting process to support FAB 11."[50] In the race to produce the Pentium chip, everything, including the health and environment of New Mexicans, is expendable.

Intel maintains a policy of not sharing important information with the public on protection of the environment. This was made all too clear when SWOP wrote a letter to the company requesting information on the use, handling, and storage of its hazardous chemicals. Intel never responded to the letter, even when Senator Jeff Bingaman sent a mild letter in support of the SWOP request. So much for good neighbors. Unfortunately Senator Bingaman was more willing to help the company get their environmental permits rather than pressure it to respond to an organization native to New Mexico.

> *The Intel plant in Rio Rancho is seventh on the list of industries in New Mexico dumping the most pollutants into the air (Right to Know).*

The addition of FAB 11 will mean huge increases in the use of chemicals and their transport through New Mexican communities, the use of scarce water supplies, increased air emissions and greater pressure on old and deteriorating infrastructures in Albuquerque's South Valley.

According to Graydon Larrabee, TI Fellow (Texas Instruments) in a speech at the International Symposium on Semiconductor Manufacturing in September, 1993, for *every single* six-inch silicon wafer processed in a FAB, the following resources are used:

    3,200 cubic feet of bulk gases
    22 cubic feet of hazardous gases
    2,275 gallons of deionized water
    20 pounds of chemicals
    285 kilowatt hours of electrical power

And for *every single* six-inch wafer, the following output is produced:

    25 pounds of sodium hydroxide
    2,840 gallons of waste water
    7 pounds of miscellaneous hazardous waste

These figures are particularly shocking when we consider that Intel wants to process 5,000 *eight inch* wafers per week when the new Rio Rancho FAB is fully operational in 1995.

In November, 1993, *with the FAB 11 only a construction site,* NMED found Intel in violation of emission standards because of production at the existing FABs. The violations were based on Intel's own figures as provided to the NMED.

Hazardous Wastes

Intel has only reported a handful of chemicals under federal "Right to Know" laws. These laws, also known as the Superfund Amendment and Reauthorization Act (SARA Title III), require all industries to report the use, handling, and storage of hazardous wastes at the facility. There are many gaps in the law. The list of chemicals to be reported to the federal government is very short and companies only have to provide information on these chemicals when they use more than a certain amount. Intel, therefore, is not required to release information on most of the 90 chemicals it uses.

To date, the company has only submitted information on the following chemicals under the Right to Know laws:

calcium hydroxide	ammonia
sodium hydroxide	hydrogen peroxide
hydrochloric acid	nitrogen
hydrofluoric acid	oxygen
chlorine	diesel fuel oil #2

As in all states, the Right to Know program in New Mexico is weak. There was no money appropriated by Congress for states to implement the program, leaving the states and local governments with a huge burden of regulation. In New Mexico, the program is supervised by the State Emergency Response Commission (SERC) and Local Emergency Planning Committees (LEPCs). The New Mexico SERC is chaired by Bill Moore, President of Rinchem, a hazardous waste storage company with storage sites in Albuquerque and Chaparral. **Rinchem is under contract with Intel to handle Intel's chemical incidents.**

The state's Office of Emergency Planning and local fire departments are responsible for implementing SARA Title III. This means, among other things, making sure that companies report hazardous waste, and storing the information. To this day, none of the "Right to Know" information in the state of

New Mexico is computerized. Boxes of files crowd fire stations throughout the state, a bureaucratic nightmare for anyone looking for information—a fire hazard in the fire station.

In 1992, the New Mexico Legislature passed a law that requires any truck hauling hazardous chemicals into New Mexico to pay a fee. This money would go to a fund to pay for the computerization of the Right to Know information and for the fire departments. The fees raise roughly $200,000 per year. Fire departments have not seen a penny of this money. Instead, the legislature diverted these funds to the state's General Fund. Neither the state's legislators nor Governor King have lifted a finger to make sure that the money is used for its original purpose.

Air

The Intel plant in Rio Rancho is seventh on the list of industries in New Mexico dumping the most pollutants into the air.[51] This puts Intel Rio Rancho in an elite class of polluters which includes the huge mining operations in the state. It also dumps much more toxins into the air than any of the California Intel factories. New Mexico allows Intel to pour 356 tons of chemicals into the air each year. New Mexico's 356 tons compares to the 10+ tons discharged annually in California's Silicon Valley in recent years. At its peak, Intel in Rio Rancho will dump more chemical gases into the air than any of the computer-chip plants and related industries in the Silicon Valley.[52]

Intel received a revised permit from the state in December of 1992 (**before the announced expansion**) to increase its emissions at Rio Rancho from 140 tons per year to 356 tons. The company admits that having the permit made a difference in choosing New Mexico for the expansion.

New Mexico's Air May Be *Too* Clean!?

The New Mexico Air Quality Bureau chief Cecilia Williams claims that Intel's 356 tons of chemicals will not violate state health standards. Because New Mexico's air is relatively clean compared to California's, air quality officials feel that Intel Rio Rancho has leeway to pollute more. This argument follows the logic of Lawrence Summers, former chief economist at the World Bank, and currently working in Clinton's administration: "I've always thought that underpopulated countries in Africa are vastly **under-polluted**, their air quality is probably vastly **inefficiently low** compared to Los Angeles or Mexico City." [53] The state also relies on Intel to develop the computer models to predict how its airborne toxics will behave. New Mexico does no independent modelling to compare with Intel's predictions. State officials cannot even be sure that Intel will stay within the 356 ton per year limit.

Currently, the most common chemicals dumped from Intel's stacks are acetone, isopropyl alcohol and ethyl-3-epoxy-

Cartoon by Kent Blair, reprinted from **Corrales Comment**.

propionate (EEP). Acetone is the main ingredient in nail polish and is linked to low blood pressure, central nervous system depression and dizziness. Isopropyl alcohol can cause eye irritation, vomiting and headaches. EEP can cause drowsiness.

The three chemicals can react with heat and sunlight to form ozone which can irritate eyes and throats and cause coughing and breathing problems.

Residents in Rio Rancho and nearby Corrales are concerned about the fumes wafting over their homes from the Intel plant. Intel insists that Corrales and other neighboring residents are smelling EEP and not acetone, although residents claim that they smell nail polish.

The complaints of the Corrales residents were routinely dismissed by the company throughout most of 1993. Then, in October, the N.M. Air Pollution Control Bureau sent Intel a formal "notice of violation." Suddenly, the company faced fines of $15,000 per day over a period of nine weeks in 1992 and for most of 1993. The violation periods coincided with the times when Intel's "hysterical neighbors" were complaining about odors and emissions. The company disputed the citations. **The basis of the Bureau's citation is the Intel record of production and the Intel calculations of emissions**. Intel's response was to "go straight to Secretary of the Environment Judith Espinosa to resolve the 'misunderstanding'" (according to a report from the *Corrales Comment*).[54]

> *Draper: "Until we get the controls on and get this new plant built up and running, we won't know exactly what those emissions are."*

The request appeared to backfire: "For some time now, state regulators and other local government officials have complained of Intel's 'arrogance' and presumptiveness."[55] More seriously, all this occurred before the planned FAB 11 has even determined the final production process, much less has a building constructed to house the operation. In January, 1994, New Mexico and Intel settled on a $40,000 payment by Intel to the New Mexico General Fund. (A bargain . . . the fines could have totalled over $2 million.) The agreement requires Intel to conduct emissions tests for volatile organic compounds on each of its stacks for the next four quarters. This is not the continuous emissions testing which should be an industry norm (this is high, high tech). But it does represent halting progress on the part of the State Air Pollution

Control Bureau. As part of the deal, of course, Intel did not admit to violating its emissions permit.[56]

In a fascinating exchange published on September 25, 1993, the *Corrales Comment* interviewed Intel spokesperson Richard Draper:[57]

> Draper: "Until we get the controls on and get this new plant built up and running, we won't know exactly what those emissions are."
> Comment: "Will new production at Intel Rio Rancho involve new air pollutants:"
> Draper: "The answer is no, we don't think so. The reason we hesitate to say a flat 'no' is because there is a new process involved, the Pentium microprocessor that will be manufactured at this plant. That is now being developed... Their corporate goal, as they progress through these generations of microprocessors, is also to upgrade their whole pollution fighting efforts toward the goal of eliminating those things that pollute. It's expensive to clean up those things that pollute and it's cheaper to cut it off at the source."
> Comment: "Is it possible that Intel's toxic or noxious emissions will concentrate in the valley, especially during thermal inversions...?"
> Draper: "Yes, it is possible. Not likely, but possible. Has site specific modeling ruled out that possibility? No."
> Comment: "Objectively, what is there about Intel's operations that Corrales residents should be concerned about. . . ?"
> Draper: "(Intel Environmental Safety Manager David) Shea says 'odor'."[58]

The company is proposing to incinerate its chemicals to eliminate the smells—a process which does not eliminate health and environmental hazards, only creates new ones. In fact the New Mexico Environment Department expressed concerns that the new process will in fact add 36 tons of toxic emissions per year. The Department is requiring a public hearing on the proposal. But the company was successful in keeping EPA out of it. Intel's Draper: "We are anxious as our neighbors . . . to get these odor abatement units turned on."[59]

The odors are a public relations problem for Intel. The odors are also a warning. Propane is an odorless gas. A chemical is added to propane so that people are aware if they are coming into contact with the fumes. Eliminating the warning signs not only avoids dealing with the real issues, but may ultimately endanger the general public even more.

And the Water

Intel's present water use is between two million gallons to three million gallons per day; it is expected to increase to an average of nine million gallons per day and may go as high as 15 million with the FAB 11. Intel has applied to the New Mexico State Engineer for permits to drill three wells. According to the application, Intel would pump 4,500 acre-feet a year (1.5 billion gallons a year). Intel would be required to purchase water rights starting at about $1,400.00 per acre foot (an acre foot is a measurement of water and is the amount of water needed to flood one acre to a depth of one-foot—about 325,852.8 gallons) or to lease water rights.

In order to obtain the water rights, Intel will likely buy them from the small farmers. Here again is a major social cost which has not been considered when measuring the Intel expansion. Most of the water rights which can be purchased are owned by indigenous families with small farms which have been passed from generation to generation. Once the water is severed from the land, the water is severed forever (Governor King: "Once the land is ruined, it's ruined forever"). The gentrification of Corrales and its neighbors will only accelerate. Intel has begun to advertise to buy water rights and is willing to pay about $1,400 to $1,500 per acre foot.

Intel intends to pump four million gallons daily from the three wells at depths of 2000 feet deep, and to purchase five million gallons from Rio Rancho Utilities Corp. Intel pays 87.2 cents per thousand gallons for water purchased from Rio

Rancho Utilities Corp.[60] Residential consumers pay a monthly minimum base charge of $7.00 for a 5/8 inch pipe or $17.92 for a one inch pipe and the rate per thousand gallons is $1.7642.

Intel is counting on the existence of an aquifer which their own experts admit may not exist. The company has drilled a test well to prove that this deep groundwater resource is real. It's like Coronado looking for gold. Bill Sheppard was asked in a recent public meeting, "What will Intel do if the water isn't there?" In Sheppard's words: "We'll have to reevaluate."

Water, water, water . . . The Rio Grande

Rio Rancho Utilities Corporation (RRUC) has applied to drill 12 new wells to pump 7.8 billion gallons annually. This would double its current supply for service to the city of Rio Rancho with a population of 35,000. Water costs to RRUC's users are among the highest in the state. As before in the AMREP advertisements of the 70's, if you want a green lawn in Rio Rancho you almost have to paint the ground.

Intel has stated it will work out an agreement with RRUC to purchase the entire nine million gallons per day should Intel's well permits be denied.

But the City of Albuquerque has applied for a permit to increase its allotment from 43 billion gallons annually to 50.5 billion.

Also, Sumitomo will locate its new expansion in Albuquerque, with approval by the Albuquerque City Council of a $67 million Industrial Revenue Bond. The company estimates it will use 1.5 million gallons a day by the year 1996. Sumitomo is a high tech company; its new plant will supply silicon wafers to Intel. Nine million gallons here, 1.5 million gallons there—all in the desert Southwest.

When the wells were first proposed, Intel trotted out

hydrologist John Shumaker, who estimated that the impact of the Intel wells on the aquifer level would be a maximum of 14.4 inches over 40 years. As questions continued to be raised, especially by Corrales, this optimism has been muted. When the Village proposed that Intel guarantee that no adverse impacts will be experienced on individual wells, the company quickly refused: "The law does not require Intel to undertake this responsibility, nor is Intel willing to undertake it," Bill Shepherd of Intel.[61]

> *In plain language: If Intel lowers the water table, drill deeper and pay for it yourself. . . . If Corrales is worried about water quality, put in a sewer system. Use the property tax. This, from the tax-exempt Lord of the Mesa.*

And Intel's thirst cannot be considered in isolation from the other requests which were flooding into the N.M. State Engineer's office. Finally, on November 19, 1993, N.M. State Engineer Eluid Martinez put all well applications on hold. The well permit applications from Intel, Rio Rancho Utilities Corporation and the City of Albuquerque threatened a cumulative withdrawal of 39,500 acre feet per year from the aquifer. In a November 19, 1993 press release Martinez stated, "(The 39,500 acre feet) corresponds to the amount of water depleted by irrigation of at least 7,600 acres and as much as 18,810 acres of farmland in the Albuquerque area. This amount of irrigated acreage may ultimately be required to be taken out of production." Action on all the requests was delayed until April 1994, while Martinez's office conducts a review of the aquifer and the water policy for the entire area. Even the *Albuquerque Journal*, hitherto almost an Intel cheerleader, editorialized on December 20, 1993: "SouthWest Organizing Project of Albuquerque and the New Mexico Environmental Law Center of Santa Fe sound an apt warning that the higher priced industrial and urban uses of water threatens small agricultural users. Particularly vulnerable are the old acequia, or community ditch, organizations of Northern New Mexico."

State Engineer Martinez' task force held its first workshop on the water issue on January 7, 1994. Intel's testimony at the hearing dropped all pretense of a low impact, "good neighbor" policy: "It may be unreasonable for any party with a shallow domestic well to expect to continue to exercise this right without continuing investment in resetting of pumps, deepening or replacement of wells, and the like . . . The needs of the community, whether municipal water supply or water supply for industry, should not be subordinated to preservation of the water level in an inefficient well that serves only an individual domestic connection." **In plain language: If Intel lowers the water table, drill deeper and pay for it yourself.**

Worried about water contamination? We quote from Intel's testimony: "Indeed, if septic tank effluent and other contamination from near the surface are more likely to move downward and enter wells because of heavier pumping from deeper parts of the aquifer, then the proper responses are to eliminate the sources of contamination by providing sewer service, and if necessary to treat contaminated water pumped from deeper wells, rather than to encourage the situation to continue by avoiding pumping groundwater from these areas." **In plain language: If Corrales is worried about water quality, put in a sewer system. Use the property tax. This, from the tax-exempt Lord of the Mesa.**

And the historic farming utilization of the acequias (community agricultural water use)? Intel: "Historic acequia associations using customary methods of surface water diversion often present a special case. If it is the case that, for example, promoting conservation would change traditional methods of surface water use that would not be economically feasible for an acequia, then the State Engineer or the Legislature may choose to adopt special rules for acequias . . . . The State Engineer should evaluate public welfare using broad criteria that reward efficient water use by applicants and protestants alike and ensure the greatest quantity of net benefits to all users of the resource while protecting the property rights of prior users." **In plain language: Grow silicon chips and not green chile.**

Indigenous peoples have farmed the Rio Grande Valley for close to 1000 years. Gary Parker, Intel's senior Vice President said "Six years after it's (FAB 11) opened, it'll look pretty old."[62]

When reasonable people step back and look at the total

costs of the Intel proposal, this is clearly no bargain. Taking thousands of acres of farmland out of production, purchasing the water from the Native American Pueblos or small indigenous farm families is a cost that in and of itself may be too high. If the water right is severed and all you can grow are subdivisions and fast food joints... an entire way of life is impacted.

The water issue must still be resolved; Intel is not assured that its applications will be approved. Corrales continues to press for some real concessions and protections from the company, or it will press ahead with its challenge. (The challenges are discussed further below in the last section on Recommendations, also see Aftermath for update.)

OFF THE FENCE by Adam Wick

Examples from INTEL sponsored WORLD CHESS. #1. The Rio Rancho Gambit

JOBS — JOBS — JOBS — CLEAN AIR — WATER — CORRALES

Cartoon by Adam Wick, reprinted from **Corrales Comment**

Wastewater

Intel currently acquires its water from Rio Rancho Utilities Corporation, a private water and sewer system that serves the entire city of Rio Rancho. Intel discharges three million gallons of wastewater per day into Albuquerque's treatment facility. Currently, wastewater flow is between 52 and 55 million gallons a day at the city plant. Intel accounts for about 6% of the daily flow and is expected to increase to about 13% of the total city capacity at current discharge rates, as a result of their expansion. Originally, Intel had requested to discharge as much as 14 million gallons a day but had to lower that amount due to city concerns that they incorporate water conservation in their plans.

Intel claims it is working to conserve two million gallons a day in order to get return flow credits and is claiming it will return between 85 and 97 percent of the water it uses. Details have not been provided. In fact, Intel is banking on three "pilot programs" to kick off its conservation scheme. The timeframe for these programs is three years.

More gambles. Intel has no other such programs at any of its other sites, and therefore has no proven technology to show that it can meet these conservation goals. What if the pilot programs do not work? Will proven technology even be in place before the Pentium plant becomes obsolete?

Intel has applied for a new waste water discharge permit, referred to as permit #2021C, to cover the FAB 11 expansion. A renewal of two existing permits #2021A and #2021B for FABs 7 and 9 is required to reflect modification for process (changes in production or volume). Approval of the three discharge permits was granted in September of 1993.

Albuquerque is expanding the treatment facility to the capacity of handling 67 million gallons per day of wastewater flow through the system. This expansion is almost at completion. An agreement between the city and Intel has been reached in which Intel would share the cost of about $10 million for the expansion of the treatment facility.

The treatment facility is located in Albuquerque's South Valley, an unincorporated area south of the city within Bernalillo County where some areas have no access to water or sewer services. The majority of South Valley residents are Chicano(a). The Kinney Brick area which is within a quarter-mile of the facility, was allocated $1.6 million for sewer lines and service in the last New Mexico State Legislative session. No construction date has been set even though an emergency clause was included in the bill. **But Intel, which is located approximately 20 miles away in its unincorporated fiefdom, has access to the treatment facility.** Intel paid about $62,000 for the sewer Utility Expansion Charge (UEC), a fee to connect to the system and to cover the cost of the users wastewater share or contribution to the system. Intel obviously got a deal with the volume it discharges daily.

Intel is provided the infrastructure at the expense of communities such as the South Valley where people have lived for over 300 years. Residents rely on individual water wells and septic systems.

Intel is provided the infrastructure at the expense of communities such as the South Valley where people have lived for over 300 years. Residents rely on individual water wells and septic systems. The area has a high water table and is susceptible to contamination from cesspools, septic tanks, leaking under ground storage tanks, Kirtland AFB and Sandia Labs, and dairies. For decades, residents have lived with facilities like the treatment plant and landfills as neighbors, yet have not been afforded access to the services. For Intel the process of accommodation and access is already in full swing.

The 1993 New Mexico State legislation allocated $12 million for water and sewer lines (projects) to be distributed equally between the North Valley and the South Valley. The South Valley project's recommendations suggest that there is a connection between Intel's expansion plans and a force main along Isleta Boulevard to provide for additional waste upon expansion completion at Intel Corporation. One public works manager with the City of Albuquerque speculated on the need to expand sewer lines due to the additional flow to the treatment plant at a cost of several million dollars which may be passed on to rate payers.

> *"If they are going to cover up something this minor, it makes you wonder what will happen when something really serious happens."*

These sewer lines are designed to accommodate the Westside and Intel's waste. Plans are being made to improve the lines to handle the additional flows which also suggests another financing of Intel expansion by tax dollars intended to provide sewer and water needs to the people of the South Valley. An initial estimate of the cost to expand these sewer lines ranges between $1 million to $1.5 million.

With the increase of water usage and wastewater, there will also be a huge increase in the use of acid neutralization. Although Intel, city, and state officials all claim that the Intel operation has been efficient and safe, there is always potential for accidental releases. What happens if the acid neutralization system fails?

Construction Accidents

Accidents have already become a regular event at Intel's new construction site. Over 80 workers were sent to the hospital after a gas leak on June 24, 1993. Two days later, Intel spokesperson Barbara Brazil announced that the fumes were from welding equipment. According to one of the workers at the site, however, "I thought at the time that that had to be a put-off. The cylinders they use don't give off that smell...I'm not anti-growth, but I am anti-exploitation. I don't like to see a company put off that little leak they had as if it were nothing. If they are going to cover up something this minor, it makes you wonder what will happen when something really serious happens."[63]

Intel, in fact, turned away the federal Environmental Protection Agency inspector, prompting the agency to send a memo to Intel chastising the company. Two days later, EPA retracted its memo, blaming the incident on "unfortunate miscommunication on both sides."

Thirteen more construction workers were injured in November of 1993 when a pipe exploded at the site.

The New Mexico Occupational Health and Safety Bureau (OHSB) is projecting 250 lost time injuries at the Intel construction site during 1994 and a one-in-four chance of a fatality. (Memo from Sam A. Rogers, OHSB Bureau Chief, to David Coss, Director, December 9, 1993) OHSB states that they will not be able to track the specific number because each of the 30 contractors at the site keeps their own logs of injuries and illnesses. The prime contractor, TDC (Technical Design and Construction), is only responsible to keep accident or injury records for their own employees. Intel is responsible to keep records for the manufacturing part of the facility only.

For instance, in the pipe explosion incident which involved a welding operation that ignited glue fumes, thirteen total employees belonging to three separate subcontractors were injured. No TDC or Intel employees were involved. This is an example of how hard it is to track specific numbers for a given site.

Intel pushes the new construction with breakneck speed.

To the company, nothing can stand in the way of its monopoly over the Pentium market. The sooner the plant is up and running, the sooner the profits start rolling in.

Meanwhile OHSB struggles to monitor thirty-four thousand industries and businesses in New Mexico with only ten full time inspectors.

Impacts on Native Nations

Although many developers and politicians have big plans for the growth of Albuquerque, the metropolitan area's expansion is held in check by its geographic location. To the east are the massive Sandia Mountains, jutting into the sky a mile above the city. To the north, the Pueblos of Sandia, San Felipe, Santo Domingo, Cochiti, Santa Ana, Zia, and Jemez continue their struggle to survive, lying between the two sprawling expansions of Albuquerque and Santa Fe. South of Albuquerque is the Pueblo of Isleta, southernmost of the Pueblos, and just downstream of the Albuquerque sewage treatment plant. Almost all of the nineteen Pueblos rely on the Rio Grande (and the aquifers and arroyos tied to it) for their existence and culture.

The Intel expansion and the state's growing addiction to electronics manufacturing pose new challenges to the survival of the Pueblo peoples. The Five Sandoval Pueblos, north of Albuquerque, whose populations number around 6,000, find themselves in numerous battles to protect their land, resources, and culture as thousands of mostly white middle-class newcomers flood the area.

Sandia and the Water

The Pueblo of Sandia's aquifer is now being pulled in every way from beneath their feet: towards the wealthy developments of the northeast heights; towards the west side for Intel and the expansion of the Silicon Mesa; towards the huge electronics factories to the south; and to the north where a factory owned by Centex pumps large amounts of water to produce wall board. Sandia, along with all of the other Pueblos may be further impacted as Intel and industrial water users continue to buy up water rights in New Mexico.

Isleta and Water Quality

Several of the Pueblos are also asserting their sovereign rights to protect their environment. The Pueblo of Isleta, Albuquerque and Bernalillo County's neighbor to the south, recently became one of the first native nations to win approval of their own water quality regulations under the federal Clean Water Act. The Act grants native nations the same status as states and allows them to set their own water quality standards.

The Petroglyphs showing continued encroachment.

The City has taken the EPA to court for approving the standards, crying that the limits set by the Pueblo for protection of the Rio Grande are too strict, and will cost Albuquerque water users as much as $250 million. The EPA won the first round in court (oh, the irony of the EPA fighting for the interests of a native nation). The City of Albuquerque is appealing and has vowed to take the fight as high as the Supreme Court.

The Pueblo of Isleta is just south of the Albuquerque sewage treatment plant, the sixth largest tributary to the Rio Grande. Recent agreements were made which set minimum flow rates, though historically, at the driest times of the year, the City's sewage effluent was the only water in the Rio Grande as it passed through Isleta. The Isleta people rely on the Rio Grande for agriculture and for ceremonial purposes. And, of course, Intel is the largest user of the Albuquerque sewage treatment plant.

The Pueblo officials stand firm in their conviction to

protect the Rio Grande. As former Isleta Governor Verna Williamson told a roomful of Public Works officials, surrounded by Native Americans with PhDs in everything from hydrology to law: "You sent us to your schools, and now we're back."

Albuquerque's all out attack against a Pueblo exercising its sovereign rights once again raises the question of priorities in development policies. For years, the City has overdeveloped its capacity at the sewage treatment plant, allowing for huge developments like Intel, 15 miles north of Albuquerque, to use up the excess capacity.

> *Former Isleta Governor Verna Williamson: "You sent us to your schools, and now we're back."*

The expansions cost Albuquerque's water users millions of dollars, paid for through Utility Expansion Fees, yet the rate-payers have no voice in the expansion decisions. The excess capacity is yet another subsidy to industry. But when the Pueblo demands that the water quality of the Rio Grande be improved, which will provide a benefit to all who rely on the river, the City is quick to attack the new standards as an "unreasonable expense."

The issues are land and water. The demands of short-sighted economic development strategies that bring massive, water-guzzling industries to a desert state will ultimately undermine the existence of indigenous people of New Mexico.

Petroglyph National Monument

Water is not the only concern to the Pueblos. The infrastructure being built to support the west side's growth is encroaching on sites that are sacred to the Pueblos. Albuquerque's West Mesa, just below Intel, is home to many petroglyphs left by the ancestors of the Pueblo people. The area is still an important site of prayer and ceremony. During the 1980's, Pueblo and Chicano people fought the National Park Service and environmentalists to prevent the site from being designated as a national park, to preserve the spiritual integrity of the site. Ultimately the federal government and environmentalists won their park, granting concessions to the native people for access and ceremonial rights.

Now the environmentalists, national park employees, and the indigenous people find themselves on the same side as they fight a proposed road through the Petroglyph National Monument. The road is being built, of course, to ease traffic congestion to and from the west side. The native people have put up a fierce resistance, and have vowed to protect the integrity of the petroglyphs. Although there are other proposed routes, the Albuquerque City Council has approved the route through the Petroglyph Park.

Public Relations

Intel has responded to community concerns with a focused public relations effort to portray a "good corporate neighbor" policy.

In early 1993, Intel hired the Hirst Company, an Albuquerque Public Relations (PR) firm, to prop up its image. Later, Intel hired on Pat Delbridge and Associates, Inc., a Canadian PR firm which lists major corporate polluters like Dow Chemical and Motorola among its clients. Delbridge has collected information on grassroots organizations and provided it to corporate clients. The company now specializes in organizing "community advisory panels," with handpicked representatives of local communities. The panels are used to present a public image of a good neighbor policy in lieu of actual "good neighbor agreements" which would commit the company to some accountability.

Look for an Intel media blitz in the nineties. Looking back at 1993, Richard Draper of Intel was quoted, "Its's been a low (point) for us (this year) because of the misperception in the community by some people of what we're about. In 1994, we'll do a better job of sharing our story."[64]

Intel has met with the Corrales Residents for Clean Air and Water (CRCAW) and seemed tantalizingly on the brink of an actual "Good Neighbor Agreement." Not to be. Intel got some PR mileage from the meetings and stepped back at the last moment. You have to understand the company. The 180 acre fiefdom on "Silicon Mesa" will remain free of the nuisance of accountability at all costs. Proposed fines can be handled politically and/or in court but production cannot be halted.

Conclusion

Clearly, the primary interest of Intel is not in seeking newer and safer alternatives to solvents. The company is mainly concerned with beating its competition and keeping its new products "clean." The health of New Mexico workers and communities will bear the consequences of the race to produce the smallest, fastest chip. In Intel's rush to make the Pentium, health and the environment are an afterthought. The tons of chemicals to be used on a daily basis at this new facility, the size of which has never been seen before in New Mexico, may be a time bomb waiting to explode.

Intel is pushing ahead at breakneck speed with the largest private construction project in New Mexico history. The production process to be used is "now being developed" (but it's already permitted). The impact on the air and water must still be determined. The company wants to reduce pollution at the source because it's "cheaper." The head of the Intel environmental effort considers "odor," not the toxics, the major concern. Intel Rio Rancho Manager Bill Sheppard: "The way you make advances in this business is by taking risks, by pushing the envelope, by experimenting with new methods. And its not just in technology. It's taking risks in everything."[65] Does anyone wonder why SWOP and the community are concerned?

Intel is relatively secure within its 180 acre island of profits. The bond issue(s) and leases have been finalized. The company will obtain the water needed, either from its own wells or from the RRUC. The Rio Rancho FAB will be the major producer of the new Pentium chip. The air permits are a done deal.

And, barring an internal catastrophe, Intel will be back for more bond money before the year 2000.

At this point, the costs of the Intel deal far outweigh the benefits to the community. This was a Faustian bargain at best—a deal with the devil. Most likely, even the economic costs will far outweigh the benefits. $250,000 per job in direct taxpayer cost is unacceptable. And, as this report and others have shown, the true cost will be much, much higher.

And when less material and even spiritual costs are included, such as the threats to the aquifer and the air basin which have sustained life for over thousands of years, this industry was a mistake for this community. Unless Intel can show the same innovative strategies in reducing their off-site impacts as shown in the company's production strategies, even AMREP may, in the long run, rue the deal.

> *Bill Sheppard: "The way you make advances in this business is by taking risks, by pushing the envelope, by experimenting with new methods. And its not just in technology. It's taking risks in everything."*

Towards A Sustainable Development Policy For Communities

Sustainable development means jobs with a future.

Corporations have no sense of mortality. Intel surely believes it will be on Silicon Mesa forever. The company was willing to pay $5 million to Sandoval County to get a new 30 year lease. Current earnings are spectacular, to put it mildly. But this industry is reducing product life cycles at an accelerated rate. And with this truly unprecedented "create and destroy" (we eat our children) industry, the company's dominance is certain to wane. The product life is three years and falling. Ironically, the

IRB subsidies do provide some assurance that the company will remain on the mesa. For Intel to walk away from tax abatement would be out of character. But when the New Mexico Investment Tax Credit is revised and when Intel comes back for more IRB money, the job issue must be directly addressed with the company. In the meantime, Intel should encourage at least some kind of worker organization which can open a dialogue regarding the job stability issue. The job issue must then be addressed openly with workers, communities, and elected officials.

Sustainable development must respect the air, land, and water.

We recognize that Intel is a driven leader in a driven industry. But this report is only another salvo in this war of the west's indigenous peoples to protect their resources. So Intel, push your engineers to develop alternative production processes that will make Intel an industry leader in more than just profits. The people of New Mexico will accept nothing less. Resource degrading and depleting industry is not wanted.

Sustainable development must respect the history and culture of the community.

New Mexico's people have survived centuries of colonization, Intel is only the latest form. Intel officials may believe that they brought civilization to the Mesa ("All that used to be up here was tumbleweeds"). But we all know that Intel is a passing phase, a destructive desert storm. Intel talks about the "Intel Culture" which has made the corporation so successful. Intel officials should be more concerned about New Mexico's culture, Arizona's culture, and Oregon's cultures.

> *Sustainable development means jobs with a future.*

Intel and the state cannot drive a road through the Petroglyphs. The Rio Grande, the lifeblood of the valley since there was a valley, must continue to flow. Three hundred fifty-six tons of toxics cannot be dropped into the community's air basin. The 180 acre "Intel property" will be subject to community control. The people will retain their land and their history. In six years, in sixty years, in six generations, the people will endure. Money, the bottom line, **will not**, finally, prevail.

Sustainable development must involve communities in the development strategy.

The behind the scenes jobmail which was played with the King Administration is an embarrassment to a company with a $2.3 billion profit in 1993! New Mexico demands more of its elected officials and of corporations operating within the state. The people will remember this in future elections. We must have representatives who will protect our interests. Of all the challenges to sustainable development, community participation is clearly the most difficult for Intel. But it is essential to the company's long term viability.

Sustainable development requires that the company pay taxes to fund the social costs of the growth.

Intel needs to drop the PR stuff and put those funds into state, county, and community coffers to pay for the schools, roads, sewer and water system expansion, landfills, police, fire protection, inspections, parks, and youth programs (the list could go on and on) which are a cost of their development.

The Intel product life spans are unprecedented. The resource demands in the West (or anywhere) are unprecedented. Nine million gallons of water a day in the desert! Six years after it's built it'll look pretty old! Companies like Intel can remain in place but must make fundamental production changes. Intel must "grow families" and not "eat babies." As this report is published, the jobmail game is being played in Arizona where the environment is just as fragile and the people just as indigenous and poor. Intel's $2.3 billion in annual profits must be invested in clean production processes which will not devastate the land, air, water, and the culture. Only if Intel can face these kinds of challenges, can show the same kind of innovation in resource management and respect for culture as it has demonstrated in profit maximization, can it prosper on Silicon Mesa.

Do not mistake the purpose of this paper. The South-West Organizing Project is not antibusiness. SWOP is not

anti-development. New Mexico needs jobs. Arizona needs jobs. Oregon needs jobs. Not Jobmail. We demand *sustainable development* which will support the community, the people and our resources.

Recommendations

In New Mexico, it's late. The horse is already out of the barn, as it were. But the lessons of this report clearly show that communities must get involved earlier and in a unified action against companies who play the "jobmail game," whether it's Intel, a high-tech clone, or any company which plays the "we're going where we get the best deal" game. These recommendations, then, address both the present condition in New Mexico and also are designed to help sister/brother communities who may face semiconductor industry jobmail now or in the future. Thus, the arguments below range from the practical to the ideal.

General

I. Organize, Organize, Organize

Intel and its competitors are both national and transnational. The SouthWest Organizing Project (SWOP) and the Southwest Network for Environmental and Economic Justice (SNEEJ) have recognized this fact and have been networking at the regional, national and international level. The Electronics Industry Good Neighbor Campaign (EIGNC) is one part of this strategy—a collaboration between technical assistance organizations and multiracial grassroots community organizations. These organizations are working together to insist that the industry act responsibly in its environmental, labor and community impacts. Community, labor and other organizations in New Mexico must also continue working together and network to demand accountability from these corporations and our public officials.

II.   Environmental, economic, social, and cultural impact studies must be completed on all major IRB projects in the future.

> *Intel must "grow families" and not "eat babies."*

Officials conducting assessments must include community groups and organizations in all notifications, and not restrict the notices to governmental officials. Grassroots community organizations represent the very people who are often most likely to be impacted by such projects. Although Sandoval County may only be the nominal owner of the "Intel Property," the $2 billion expansion on Silicon Mesa is, ultimately, being undertaken under the county's name. A governmental entity cannot be allowed to indirectly fund a $2 billion project without a full analysis of the environmental, social, and economic impacts. Sandoval County can make such a commitment. Then when Intel comes back for more IRBs, the full impacts of the development will be opened for community analysis, comment and response. The requirement for an impact analysis will cure many of the problems which are currently associated with the FAB 11 expansion. Materials must be in the language spoken by local communities.

III.   Tax abatement policies for development must be fully reviewed and serious and independent cost/benefit analysis must be applied.

The New Mexico Economic Development Department has no idea of the costs of the Intel deal. The New Mexico Bureau of Business Research makes the very same point. New Mexico legislators likely had no idea about the full impact of the NMITC. A high level task force, composed of community representatives, legislators, business representatives and educators, must be created to review and analyze the New Mexico tax incentives for business. The current policy passes on the cost of development to all New Mexicans, including New Mexico companies.

IV.   Intel, and other high-tech companies, must adopt the "Community Bill of Rights," the "Principles of Environmental Justice," and the "Silicon Principles," which can be found in the appendices of this book.

These principles can guide both communities and companies toward what is, finally, a common goal: Community Sustainable Development.

SWOP crashes "Intel Party" in Albuquerque, NM, 1/26/94, "No Gracias Intel! Super Profits, Super Toxic Pollution. Real New Mexicans Pay Taxes!"

Air

I.   Intel and other high-tech semiconductor companies must implement continuous emissions monitoring. Their neighbors must know what invisible brew is emitting out of their immaculate stacks.

II.   The NMED must be funded to provide for independent analysis of the air emissions problems. To date, all the figures are Intel's. The NMED must develop its own data and independent analysis.

III.   Air quality monitoring must be implemented within Corrales and other affected communities. Modeling of the air basin, the impact of winter inversions and the "Albuquerque Box" syndrome (you lift off, you float, you land close to the take off point) known to participants of the International Balloon Fiesta, must be documented. The health of New Mexican citizens is at stake.

IV.   The giant "odor eater" incinerator is no answer. We reject this "solution." It's not the odor, stupid. The smokeless stacks on the mesa will not convince us. Until the air in Corrales and Rio Rancho is monitored by NMED, the communities will not be safe. What we can't smell, can hurt us. It is the emissions, stupid.

Water

I.   Intel can't own the water; Intel can buy the water from Rio Rancho Utilities Corporation (RRUC).

II.   Make the company purchase the water so that there will be an economic incentive (money, remember, the bottom line) to conserve. Subsidies to big industrial users by publicly regulated utilities must cease.

III.   Intel must commit to conservation. This lesson of the water hearings must be clear and unambiguous. Deny Intel the permits.

IV.   A community-defined, comprehensive water use policy must be developed and implemented. Industry must commit and invest in a 100% water reuse conservation plan.

V.   Water quality regulations must be strengthened and enforced. Waste water discharges must be subject to "Best Available Control Technology."

VI.   The City of Albuquerque must drop opposition to Isleta Water Quality Standards and concentrate on compliance. Spend the money on cleanup and not attorney's fees. (On April 10, 1994, Mayor Chavez and Governor Lucero of Isleta announced a tentative settlement.)

VII.   Even if Intel's well permits are denied, the requests of the RRUC and of Albuquerque will impact agriculture. These impacts must be addressed by the State Engineer. Funds must be set-aside to protect water uses which are a part of the Rio Grande Valley's culture.

Emergency Response

I.   Intel must disclose **all** chemicals used at the FABs.

II.   Intel must tell us where the chemicals are going and when, in the disposal process. If Intel toxic wastes are trucked to Arizona, New Mexico is still a party to this game. And all communities on the trucking routes must be aware of the routes and the schedules.

III.   Intel must publish Worst Case Scenarios for all of the chemicals which it uses.

IV.   Intel must provide full and immediate disclosure of releases and accidents. The chemicals in use at the FABs are serious and can impact public health.

V.   Government (this is another cost on citizens) must be prepared for Worst Case Scenarios in the event of a spill or incident, with a response plan.

VI.   Intel should provide for open inspections of Intel facilities by community representatives.

Jobs

I.   Open, binding commitment goals for employment of New Mexicans must be established. Reports on affirmative hiring goals and accomplishments must be public and routine. Intel can't receive a subsidy of $250,000 (over five years) per job and not report who is getting them.

II.   Affirmative Action in hiring must be in place for all job levels.

III.   Benefits packages must include quality medical/ dental benefits at reasonable cost. Child care services must be provided by the company to assure equal access to jobs for all.

Worker Safety

I.   New Mexico's Occupational Health and Safety Bureau and the Construction Industries Division must make a thorough review of the Intel FABs. Regular inspections of Intel and other semiconductor companies must be performed to assure that workers are fully informed of the workplace hazards and that safety policies address the serious dangers inherent in this business. Not only incident situations, but, also the nature of long-term chemical exposure, are essential to worker health and safety.

II.   Intel must take affirmative steps with a worker awareness campaign, which goes beyond the usual industry rhetoric and addresses the real dangers of exposure to the chemicals which are the basis of this "clean industry."

III. The public (and the Landlord, Sandoval County) must be provided all documents relating to environmental, and worker safety, as required under OSHA and EPA rules on chemical accident prevention.

IV. Let the workers organize. The anti-union, anti-worker stance must change to address the realities of the workplace. Intel is in New Mexico because our workers are productive. The tax breaks and environmental concessions are just icing on the cake. New Mexican workers need good pay, jobs with a future, and a safe workplace. Workers should be allowed to develop a worker's "Bill of Rights."

V. The New Mexico Worker Compensation Law must be amended to assure that workers are on equal footing with companies in claims cases, and to assure that workers are fully compensated for on-the-job injuries.

VI. High-tech companies, like Intel, must establish a fund to compensate and treat workers who are damaged by long-term exposure to chemicals within their workplace. Worker's health must be protected even after companies leave the area or close. The long-term costs of cases like GTE Lenkurt are imposed on the workers, their families, and on the society who must support them, after the companies are gone.

Silicon Valley, California toxic tour press conference by Electronics Industry Good Neighbor Campaign   *photo by Grant Atwell*

Government

I. The New Mexico Investment Tax Credit must be redirected toward small New Mexican businesses.

II   IRBs must be used for sustainable development and their term must recognize the product's life-span. If a chip has an expected economic life of five years, set up the IRB and the lease for five years.

III.   The state, county, and community governments must make decisions about development in the open and in consultation with community organizations, religious groups, and labor representatives.

IV.   IRBs must be limited to companies with specific plans to implement pollution prevention and resource conservation.

V.   Local governments must restrict IRBs to New Mexico based corporations, for long-term New Mexico jobs.

VI.   Protections must be written into the Intel lease (in case the Lord of the Chip gets beheaded or the gun goes off) so that New Mexico can retrain workers who lose jobs in a massive shutdown.

VII. Stop the road through the Petroglyph National Monument. The Albuquerque west side and companies like Intel that are fueling its growth must pay the cost of a route that respects Native Nations sovereignty and beliefs.

VIII. Elected officials should meet with community groups from other cities where an industry is moving from, to better understand the history and track record of the industry.

IX.   Funding should be secured from the industry for community participation.

> *The genie is out of the bottle, or as New Mexico ex-Governor Bruce King might put it, "the Pandoras are out of the box."*

Aftermath
March 1, 1995

This supplement updates *INTEL INSIDE NEW MEXICO* from May 3, 1994 to March 1995. As predicted, the company remains financially healthy and has obtained the permits and infrastructure needed for the New Mexico expansion. At the same time, the people of the state are looking much more critically at the bargain. The Industrial Revenue Bonds (IRBs) and leases have another 29 years to run, so the story won't be over for a while.

Jobmail, Intel's Model For Corporate Welfare: The "Intel Ideal Incentive Matrix"

The May 3 report: "Given the success with FAB 11, this process is certain to be used again. The calculating, cynical process. . . perfected in the FAB 11 sweepstakes, appears to set a new low in "corporate jobmail." The SouthWest Organizing Project has now obtained copies of both the California and the New Mexico final proposals to Intel; we now know just how low the state went.

Each community which Intel flirted with was given a "FAB 11 'IDEAL' INCENTIVE MATRIX" which outlined the company's perfect package. And each community went through a first process, further demands and negotiations, and then a final submission. The final October 12, 1992 "New Mexico Matrix" is attached as an appendix to this supplement (see Appendix IV). The "Intel Ideal Incentive" is what you would expect: no taxes, no permits, labor subsidies, all power and infrastructure at no expense to Intel. And New Mexico did literally run away with the FAB 11 sweepstakes. New Mexico included 17 Intel employees on their proposal team, including David Shea and Barbara Brazil.

Industry expert Kathleen Wiegner was already reporting in June, 1992 that Intel was expanding the New Mexico FAB for the Pentium production. But while industry insiders figured

Intel had already decided on New Mexico, the company pursued the "Intel Ideal Incentive" to bleed a few more concessions from the state.

Intel Inside Arizona and Oregon

In July, 1994, Intel pronounced Chandler, Arizona the latest "jobmail sweepstakes" winner, with an incentive package estimated by Gary Tredway of *The Current* of Phoenix, Arizona at $169 million, about $89,000 per job. The *San José Mercury News* reported on California's second loss in 15 months: "California is pitted ruthlessly against other states in blind bidding....It doesn't matter whether the subject, Intel, has more than $4 billion in the bank....No doubt about it, Intel is a tough bargainer, says Julie Wright, the head of the California Trade and Commerce Agency in Sacramento. 'Intel is very experienced at this process,' she says. 'They know what they want.'"[66]

In August 1994, Intel played its game again. In Oregon Intel requested the Washington County commissioners to approve a property tax exemption for the company on a plant expansion proposed in Hillsboro, Oregon. The proposal would exempt all but the first $100 million of taxable value from the property tax. Washington County was the first county in Oregon to implement a change in state law which allows such an exemption. On Monday, August 15, 1994, the eve of the vote, Intel officials were explicit in their jobmail threats.

Keith L. Thompson, Intel's top executive in Oregon made the threat clear: "We can expand in a number of locations. If we don't have the incentive for Oregon, then Oregon is not competitive." Bob French, Intel's government affairs manager was smoother, at least for public consumption: "We feel confident that the Washington County commissioners recognize the positive economic impact of further Intel expansion." The commissioners voted 5 to 0 for the tax exemption on August 16.[67]

Jobmail may be too polite a term for the Intel process.

The New Mexico Incentive Package

The publication of *INTEL INSIDE NEW MEXICO* created an uproar with its documentation that the tax and other incentives were much greater than announced by the state. New Mexico spokespersons had taken credit for the incentive package when the deal was announced on April 1, 1993. The state stuck with its original estimate of $114 million of incentives. John McKean, Governor King's Press Secretary, replied that Intel got only the same kind of incentives that any other company could get.[68]

> *Keith L. Thompson, Intel's top executive in Oregon made the threat clear: "We can expand in a number of locations. If we don't have the incentive for Oregon, then Oregon is not competitive."*

The October 12, 1992 New Mexico proposal has a different spin. The proposal cites "customized tax and utility legislation." In the proposal the actual dollar value of the New Mexico incentives is estimated at $304 million! And the $304 million package, it must be noted, is based on a $1 billion expansion and on 18 years of IRB tax shelter, rather than the $2 billion actual expansion with a full 30 years of the IRB tax shelter. The additional savings accruing to Intel from the change to allow companies to use a double weighted sales factor to reduce their state income tax liability was also not yet in place. The New Mexico proposal clearly shows that the double weighting which benefits companies with the majority of their sales out of state was an "Intel Amendment."

On July 28, 1994, *Albuquerque Tribune* reporter Tony Davis published a story which showed that the *INTEL INSIDE NEW MEXICO* numbers were much more accurate than the numbers the state was providing (see Appendix VII, page 118). Davis concluded that the total **net loss** to New Mexico over 30 years would be $2 million. The analysis used present value calculations, discounting both the tax breaks and the tax revenues which are estimated to be generated over 30 years. Jonathon Krebs, New Mexico's Economic Development Department secretary responded: "I don't frankly understand present value calculations." Although Mr. Davis's report was an important contribution to the Intel story, there remains a serious flaw in the analysis. Intel is not "your father's Oldsmobile company." Intel will require an additional infusion of Industrial Revenue Bonds in six years if the Rio Rancho FABs are to remain in operation. With the cost of a new FAB quadrupling every five or six years, the IRBs will greatly increase the tax forgiveness. The state and local governments will never catch up with tax revenues from this company.

If anyone doubts the exponential increase in IRBs look at the history:

Year	Amount
1980	$1 million
1981	$29 million
1984	$60 million
1984	$3.5 million
1989	$240 million
1990	$500 million
1993	$2 billion
TOTAL:	**$2.8 BILLION**

What About The Jobs?

The original Intel proposal promised 1000 jobs for New Mexicans. With the publication of *INTEL INSIDE NEW MEXICO*, the company immediately increased the number of promised jobs to 2000. From the news releases trumpeting the FAB 11 plum, most of these jobs could be expected to go to New Mexicans. But recent revelations and the New Mexico incentive proposal of October 1992, show that most of these jobs will go to "new New Mexicans," i.e., workers from out of state, who will become New Mexicans only after they relocate here.

One section documents the "Human Resources Incentives" which New Mexico, in conjunction with Real Estate Corporation of America, Albuquerque Title Company, Delta Airlines, Commonwealth-United Mortgage, and New Mexico utility companies assembled to reduce the cost of relocation for the new FAB 11 employees. The total value of the incentives (which includes a $300,000 value for the state funded TVI-Intel facility) is $4.2 million. The explicit assumption in the accumulation of these incentives is that 524 (52%) of the 1000 new jobs will go to people from outside New Mexico! Intel gets tax breaks of from $114,000 to $300,000 per employee and over one-half of these employees will **not** be local hires. With the stated increase in the number of jobs to 2400, local workers will be an even smaller percentage of the hires. On September 9, 1994 Intel made a preemptive strike to mute criticism. Intel's Yvonne Lenbergs told Paul Logan of the *Albuquerque Journal* that "Intel has 'to a large degree' nearly exhausted all the instate qualified workers at the technical level."[69]

Intel, with much fanfare and press coverage, has also promised 100 scholarships of $3000 each to students at TVI and the Southwest Indian Polytechnic Institute (SIPI). But the company will even make a profit on the scholarships. Successful scholarship recipients can expect to be put on the Intel payroll at approximately $22,500 per year. Intel can then use New Mexico Job Training Funds to pay 1/2 of the employees first six months wages. The state, in other words, will pay Intel $5625 for the successfully placed scholarship student. Intel nets $2625 per hired student.

Intel steadfastly refuses to provide any detailed information on its workforce and hiring practices, not just to SWOP and to SNEEJ, but to the New Mexico labor department. When it gives out a number such as Barbara Brazil's 60% of new hires in 1993 were New Mexicans, there is no check available to anybody. EIGNC has written Intel Gordon Moore for answers but does not expect a quick or forthcoming response.

Note: In January, 1995, Intel announced that 61% of FAB 11 hirees listed a New Mexico address on their application. These figures, as with so many which are released by the company, have not been independently confirmed. It is logical to assume that some of the "new hires" came from the other New Mexico FABs, and may have originally come from out of state. Hopefully, the upcoming SNEEJ/SWOP/Intel meetings will answer our concerns.

The Revolving Door

But there is one well-documented case of a New Mexico hire by the company. On June 22, 1994, Bill Garcia, head of the New Mexico Economic Development Department and point man on the Intel Incentive Team, left state government to "manage Intel's lobbying and other government relations at. . .a 'substantial' boost in salary."[70] Jeanne Gauna of SWOP responded that the action "reeks of conflict of interest. It's now obvious that Garcia was the chip-making giant's servant all along."[71] How correct was Ms. Gauna? Try this from a letter dated March 31, 1992 which Mr. Garcia wrote on New Mexico letterhead to Mr. Michael R. Splinter, VP of Intel: "It has been my pleasure to work with Intel in the past and it will certainly be my pleasure to work with Intel in the future as well. . . .I feel the best role for the New Mexico Economic Development Department is to be a tool to be used by companies such as Intel. . . .It was our pleasure to push recent legislation to allow the continuation of the Investment Tax Credit, a tool Intel has used extensively in the past and will have available in the future, should your site selection team recommend Rio Rancho. This business development legislation is unique in the nation and certainly one which has at least helped Intel reach the profitable position the company is in today." (see Appendix V, page 114)

SWOP certainly concurs that the Investment Tax credit is a unique tool. Intel collects the credit by retaining employee state income taxes. And protecting this unusual and misguided incentive from the New Mexico legislature is likely the key reason that Intel hired Mr. Garcia.

In October 1994, it was announced that David Scott, chief staffer at Albuquerque Economic Development, Inc. who worked to bring Sumitomo Sitex to Albuquerque with an incentive package similar to Intel's, had been hired by Sumitomo to be the company's top administrative officer in New Mexico.

New Mexicans can at least count on one local hire per proposal.

It's The Odor, Stupid!

In August, 1994, with great fanfare, Intel turned on the three thermal oxidizers, the giant "odor-eaters" on silicon mesa, to run continuously. Barbara Rockwell of Corrales Residents for Clean Air and Water (CRCAW) was on hand to help pull the switch. CRCAW had negotiated with Intel representatives to assure that the odor-eaters operated continuously. Their participation and blessing represented a major public relations victory for the company. Intel had, of course, firmly committed to solving the "odor problem" 12 months earlier.

In August, Intel received the air permit promised by New Mexico and narrowly averted an Environmental Protection Agency (EPA) review of the entire proposal under federal regulations for "prevention of significant deterioration" (PSD). The PSD process could have delayed the plant by up to 12 months. Intel's gyrations to avoid the EPA review are standard operating policy for the company and were documented in the *Corrales Comment*. The FAB 11 expansion required 12 new 1,250 horsepower boilers in its energy center for heating and cooling. The boilers were initially calculated to emit **433** tons of emissions annually. The EPA entered the case because they determined that emissions in excess of **250** tons per year will trigger a PSD review. "Intel scurried to rectify that apparent discrepancy, and corrected its submissions to show that the emission level would be less than 215 tons, not 433. . . ."[72]

Then on May 12, 1994, Stanley Meiburg, Director of the Dallas EPA

> *Bill Garcia, head of the New Mexico Economic Development Department and point man on the Intel Incentive Team, left state government to "manage Intel's lobbying and other government relations at. . .a 'substantial' boost in salary."*

Regional Air, Pesticides and Toxics Division wrote the NMED: "The proposed combination of boilers, standing alone, qualify as a listed source, and would potentially emit a pollutant at 100 tons per year or more. Therefore, the combination of boilers are a major source and are subject to PSD review."[73] Back to the drawing board for Intel. The company immediately set its technicians to work and recalculated that the boilers "would recalculated that the boilers "would not, after all, exceed the 100 ton-a-year threshold."[74] Intel ultimately altered its permit to show the reduced emissions and the NMED and the Dallas EPA accepted the revised permit. Once EPA accepted the revised Intel figures, the air permits were a done deal. This is vintage Intel. The company goes on record. If their scientific calculations need changing to forward their goals, with no embarrassment, the numbers are changed. And the company continues to assert that only Intel can provide data which is accurate.

But continuous emissions monitoring (CEM) is too expensive. At the June 10, 1994 air hearings, Gail Hoffnagle, an air quality expert under contract with Intel, testified that CEM of the boilers would cost $1.5 million for the first year. One of Intel's scientific formulas will be as accurate.[75] At every step of the process, Intel has refused to consider CEM, which is the only real way to assure proper operation of equipment and to assure that air standards are met. $1.5 million for CEM is too much money; $11 million for the odor-eaters is no problem.

And The Water

Intel's water request served as a wake-up call for this desert state. Intel plans to use up to nine million gallons of water a day when FAB 11 is in full operation. Five million gallons would come from the Rio Rancho Utility Corporation. Four million gallons were to be pumped by Intel from its own wells at a cost of $.25 per 1000 gallons.

> *$1.5 million for CEM is too much money; $11 million for the odor-eaters is no problem.*

New Mexico State Engineer Eluid Martinez ruled on June 10, 1994, that Intel can pump approximately 2.9 million gallons per day from their wells, 72% of their request. Martinez also required the company to set up a well monitoring system in Corrales to report on impairment. And Intel was required to design and implement a water reduction/conservation plan. This type of a compromise was predicted in *Intel Inside*. By pushing for the hearings and raising the issue, however, groups like SWOP and SNEEJ have definitely created a new consciousness, even among the go-go pro development forces. Water in New Mexico is a life source. While Intel has won the battle, the assurance which New Mexico gave Intel in October 1992 won't be given again.

State Engineer Eluid Martinez issued new regulations in September on the practice of water rights "dedications" which allow someone to buy and withdraw water rights in exchange for permission to use an equal amount of water elsewhere. The new regulations are a response to a water task force report which was created as part of the response to the Intel water request.[76]

Albuquerque Mayor Martin Chavez has made water policy a major issue for his administration. On June 2, 1994, he announced that he refused to guarantee a California semiconductor company one million gallons a day, which the company demanded in order to seriously consider Albuquerque for an expansion.[77]

The City of Rio Rancho has filed a condemnation suit to take over the RRUC. The two parties are in court. Regardless of the outcome, short-term rate relief for the citizens is not anticipated, even by the City.

"WHICH INTEL GOVERNOR DO YOU WANT?"
1994 Green party candidate Roberto Mondragón

The November 1994 election was Bruce King's "last hurrah." The Intel giveaway, coupled with his abandonment of Chicano(a) and Native American issues, led to an unprecedented challenge in the primary, which King barely won. Shortly after the primary, the Green Party nominated Roberto Mondragón as it's candidate for Governor. Mr. Mondragón ran a strong third party campaign raising exactly these issues.

The Republican winner and chief beneficiary of King's misguided policies is Gary Johnson, New Mexico's new Governor. Through Mr. Johnson's Big J Enterprises, Johnson is under contract to Intel. A report in the September 4, 1994 *Albuquerque Journal* documents the connection. "Big J works primarily at Intel's Rio Rancho plant. Johnson won't say what percentage of Big J's work comes from Intel. 'We'd like to make it less important to us, but it can't be insignificant, because it is the largest construction project in North America, and we have a big foot in the door,' said Johnson."[78]

Mr. Johnson seems to have the Intel mentality as well as their contracts. The connection began in the mid-1980's. Art Stout, Intel's senior projects manager: "I was evaluating how to handle the construction work force, because we. . .didn't want to be in the construction business. And we didn't want the administrative headache of dealing with a work force that would be continually expanding and contracting." For Johnson, the Intel deal saved his company: "People shouldn't fault us for that. We do the inglorious hiring and firing. We deal with the administration of upsizing and downsizing a work force of several hundred."[79] As the campaign heated up, former and decidedly disgruntled workers came out with "charges of firing of whistleblowers, laying off workers just before they're eligible for benefits and quashing union activity."[80] A rigged test of a sewer line at the Intel plant was confirmed, the line was retested and passed.

A political novice, Johnson ran a standard pro-growth, cut taxes, don't ask for specifics campaign. In a cynical twist to SWOP's *Reclame Nuevo México* (Take Back New Mexico) campaign, Mr. Johnson's campaign slogan was "Win Back New Mexico."

Intel must have watched the campaign with bemused detachment. Mr. Johnson won as New Mexico's governor for the next four years. One Intel governor was replaced by another.

Intel Answers To No One

Intel has still not responded to the *INTEL INSIDE NEW MEXICO* report. The company has not been responsive to stockholders, either, if the stockholders' questions make it uncomfortable. Stephen Viederman (author of the forward to this report) is president of the Jessie Smith Noyes Foundation, which owns 3000 share of Intel stock. Mr. Viederman attended the Intel stockholder meeting in May 1994. Mr. Viederman asked Andrew Grove the same questions. No real response to date, but Viederman has received great praise in progressive foundation circles for the effort. Timothy H. Smith, director of the Interfaith Center on Corporate Responsibility stated: "The Noyes Foundation's appearance at the Intel stockholder's meeting is the first time a foundation executive ever has taken such a step and raised tough questions with corporate management. Noyes is on the cutting edge of a new trend."[81] Henry Goldstein, writing in *The Chronicle of Philanthropy,* July 12, 1994: "The Noyes Foundation has displayed creative hell-raising at its best. . . .What began as a single effort might well be organized and orchestrated to deal with a range of social issues. That would be a useful activity for the Council on Foundations. . . .If Mr. Viederman's ploy goes anywhere, it may be time to buy."[82]

Roberto Mondragón, third party candidate for governor.

Mr. Viederman remains committed to our efforts. Intel remained silent for quite a while. Intel has still not met with SWOP, SNEEJ, or EIGNC, however, as this document goes to press (March, 1995), SWOP and Intel have agreed to meet with each other and are negotiating over the process.

$$INTEL$$

How's the company doing? Right now, pretty well. Intel exploits its monopoly hold on the computer chip market just as it exploits its economic strength in dealing with communities. 1993 pre-tax profits were reported at $3.392 billion on revenues on only $8.762 billion. After tax earnings were $2.295 billion. Earnings reported in July 1994 had increased 13% over 1993. In October, 1994 another record quarter. Andrew Grove drew $2.4 million in salary and other cash compensations in 1993. Four other officers received over $1 million each. Special officers such as Grove and other high level officials (how about Bill Garcia?) also have special stock purchase and options rights, which guarantee them a profit. The company has an estimated $4 billion in cash and short term assets in the bank.

> *Intel could afford to pay some taxes.*

Intel could afford to pay some taxes.

Intel, The Good Corporate Neighbor

On Thursday, May 20, 1994, an Intel construction worker hit a transmission line and knocked out power to Intel and about 9000 other customers in Rio Rancho. A second outage occurred later in the day. Intel offered no apologies to Rio Rancho residents. Don Hutchinson, Intel central operations manager decried the outages as "about the worst thing that's ever happened to us in a while." The unexpected shutdowns were "possibly the worst ever" with losses "in the millions of dollars."[83]

Poor Intel. What was not disclosed by the company was that the company was immediately planning to collect on an insurance policy which Intel had purchased to cover losses in just such a case of a construction accident. The company preferred to bask in the sympathy of the community by trumpeting its losses. Other businesses and private citizens in the affected area also suffered losses and could have benefited from the knowledge of the insurance policy.

Intel's Public Relations firms are working at the same breakneck speed as the FAB workers.

The Workers

Since publication of *INTEL INSIDE NEW MEXICO,* many current and former Intel employees have contacted SWOP regarding their conditions. The employee reports paint a disturbing picture of a demanding, paranoid Lord of the Mesa, who commands absolute loyalty to the Intel family.

In August, 1994, Intel opened a huge, 750 seat cafeteria inside Intel. As reported in the *Albuquerque Tribune*, the eatery is an effort to keep workers inside during breaks.[84] Local restaurant owners were not impressed.

A Pentium Postscript[85]

If the reader has stuck with us so far, the Pentium controversy is just a classic case of the Intel mentality. No surprises here. In July, 1994, the company discovered that the Pentium chip had a flaw. Intel continued to produce the chip and told no one; indeed the company continued to produce the flawed chip while working on a fix. In late October 1994, Thomas Nicely, a professor at Lynchburg College in Virginia, put a letter on the Internet about the flaw, which could produce a mathematical error. The company stonewalled and stonewalled. Intel insisted that the flaw was so minor as to cause no problems for most users. The company agreed to replace a Pentium if a customer demonstrated a failure. The customer would have to pay for the service charge of replacement.

Then IBM jumped in and agreed to replace any Pentium chip at no charge. Soon, IBM raised the stakes and stopped shipping IBM computers powered by the Pentium until the problem was resolved. The Intel stock began to drop. On December 20, 1994, Gartner Group, a major technology consulting firm, advised clients to hold off on buying large stocks of Pentium powered PCs.

Cash and profits became the issue. And Intel got the message. On December 20, 1994, Intel reversed its policy and agreed to replace chips at no cost and no questions asked. CEO Grove: "To some people, this policy seemed arrogant and uncaring. We were motivated by the belief that replacement is simply unnecessary for most people. We still feel that way, but we are changing our policy because we want there to be no doubt that we stand behind this product."[86]

The Pentium fiasco is no isolated incident. Paul Cassel, an independent computer scientist who writes a column for the *Albuquerque Journal's Business Outlook* and for New Mexico's *Computer Scene,* has had enough. In the March 1995 *Computer Scene* article "I'm Fed Up!," he carried the reader through the Pentium debacle and then to a problem he had with an Intel motherboard, which had defective chipsets. Intel's tech support sent him to the Intel Bulletin Board Systems (BBS) where the files and software to fix the chipsets were available. Nothing there, nothing on Intel's CompuServe forum or Internet server. Cassel by chance was led to the files. He went back to Intel and their BBS. In fact the corrective files were on Intel's BBS but in a "private section" and not available to the general public. Mr. Cassel let his frustration out: "In a few short months I've seen too many facts of Intel I don't like. The company is miserable to deal with, and it doesn't have any spectacular technology which can make misery tolerable . . . I'll make an effort to find non-Intel solutions. These might come from AMD, Cyrix, Apple or IBM— any of which I'd prefer to deal with over Intel. I'm not alone. Intel has got to realize that no ship is so large that it can't be sunk . . . 'Intel on the Inside' now means that you've bought expensive, indifferent technology vended by a disdainful, deceitful manu-facturer."[87]

Reclame Nuevo México/Take Back New Mexico

Reproduced on the opposite page is a graphic from the submission of the 1992 New Mexico proposal to Intel. If Intel continues to see its role as a Lord of the Mesa, with profits as the **only** motive, the company will indeed be only a "brief desert storm."

Sustainable development means respect for air, water, land, people and culture. The people of New Mexico demand community sustainable development.

Footnotes

1. "Proceedings, The First National People of Color Environmental Leadership Summit," Washington, D.C., October, 1991
2. Eastman, Carruthers & Liefer, "Contrasting Attitudes Toward Land in New Mexico," *New Mexico Business*, March 1971
3. New Mexico People and Energy, "Land Subdivision: The Selling of New Mexico," *Power Structure Report No. 6,* Fall 1980
4. Kathleen Wiegner, "The Empire Strikes Back," *Upside,* June 1992
5. SWOP wants to thank Lenny Siegel of the Pacific Studies Center for the basic industry information for this section. Mr. Siegel's report was then supplemented by a fascinating article by Kathleen Wiegner in *Upside,* June 1992, "The Empire Strikes Back." Ms. Wiegner is an industry

Final Summary

What Intel Can Expect By Placing Fab 11 In New Mexico

and Intel expert and her insight into the 'Intel Mentality' is enlightening.

6. "The Empire Strikes Back," *op. cit.*
7. *ibid.*
8. Christopher Barr, "Pentium Killers," *PC Magazine*, January 11, 1994
9. Paul Logan, "Intel Considering Research Department," *Albuquerque Journal, Metro Plus*, December 4, 1993
10. "The Empire Strikes Back," *op. cit.*
11. Gordon E. Moore, "Streamlined EPA Rules Crucial For High-Tech Success," *San Francisco Chronicle*, October 12, 1992
12. Rich Karlgaard, "Michael Slater," *Upside*, June 1992
13. "The Empire Strikes Back," *op. cit.*
14. The Silicon Valley Toxics Coalition (SVTC) provided the basic information on Intel in California. Working together and sharing information is essential for community groups to understand transnational companies like Intel.
15. Jeff Radford, "Why Tighter Air Standards Are Needed For Intel Plant," *Corrales Comment*, September 25, 1993
16. Sources: *EPA and RWQCB Fact Sheets on the Intel Sites*, April, 1991
17. Sources: *EPA Toxics Release Inventory, 1987-1992*, and *BAAQMD Toxic Air Contaminant Emissions Inventory for 1990-91*
18. Rebecca Smith, "Intel Bucks Trend With 250 New Factory Jobs," *San José Mercury News*, October 7, 1992
19. Jim Carrier, "Toxic Puzzle: GTE Plant Left Trail of Bills, Death," *The Denver Post*, June 13, 1993
20. SouthWest Organizing Project, *Report On The Interfaith Hearings On Toxic Poisoning In Communities Of Color*, April 3, 1993
21. Maxcy Rosenau, "Diseases Associated With Exposure To Chemical Substances: Organic Compounds," *Public Health and Preventive Medicine*, 12th Edition, J.M. Last, Editor, 1986
22. Karen Hossfeld, "Why Aren't High-Tech Workers Organized?," *Common Interests: Women Organizing In Global Electronics*, Women Working Worldwide, 1991
23. Shonda Novak, "Intel Asked Gore For Reassurance," *Albuquerque Tribune*, April 5, 1993
24. "Land Subdivision: The Selling Of New Mexico," *op. cit.*

25. *ibid.*
26. Intel alone among New Mexico manufacturing companies even has a branch of the Albuquerque Technical Vocational Institute located on the 180 acres. This state subsidized school is open to the public, assuming that the retirees of Rio Rancho are interested in the Winter 1994 offerings: four courses in "Electronics Fundamentals," six courses in "Semiconductor Devices," two courses on "Digital Circuits," three on "Introduction to Microcomputers." . . .Oh yes, there are three English courses offered.
27. Rebecca Smith/Thomas Farragher, "Why California Lost Bid For Intel Plant," *San José Mercury News*, April 4, 1993
28. SWOP has now obtained the "Intel Ideal Incentive Matrix" used in the New Mexico process, see Appendix IV, p. 106
29. Rebecca Smith, "Beauty Of Landing Intel May Only Be Skin Deep," *San José Mercury News*, February 22, 1993
30. *ibid.*
31. Bureau of Business and Economic Research of the University of New Mexico, New Mexico Tax Study, Phase II Report, December 17, 1993
32. The tax breaks in place on the current property keep going and going and going, like the TV rabbit. Mr. David Marsing, Intel Spokesperson at the August 16, 1993 County Commission meeting, estimated that the cost of new projects was quadrupling with each new chip on the block. If he is correct, we can anticipate a $6 Billion project on the books five years later. Remember, this began with a $40 million deal in 1980. The total in tax abatement for property taxes for the next five years would be over $300 million. Sales taxes are levied at 5.5% in the County. 5.5% X $5 Billion is $275 million. In a second expansion phase, the projected property abatement tax and sale tax loss would be $675 million. If you project further, the numbers just run away from you. But if Intel remains viable, the numbers will be there and the lost revenues become staggering.
33. Intel has written SWOP that the company will pay $22 million in income tax to New Mexico in 1994. In this case the actual tax savings to Intel are $7 million in 1994 alone.
34. Isabel Sanchez, "Area's New Trade Zone May Aid Economy," *Albuquerque Journal Metro Plus*, August 21, 1993

35. Bob Hagan, "Workers's Comp Reforms Pay Off," Business Outlook *Albuquerque Journal*, July 19, 1993
36. *ibid.*
37. *ibid.*
38. When the original *Intel Inside New Mexico* report was released in May 1994, the documentation of the tax incentives created much controversy and comment. Intel immediately increased the job creation from 1000 jobs to over 2000. See page 81 of the *Aftermath* for an update.
39. We have calculated $250 million over five years in tax abatement and various subsidies. The total projected loss over the 30 year life of the project will be astronomical. Remember, the company states that the cost of a FAB is quadrupling. If so, most of the $250 million will quadruple right along with the cost. If this holds, the tax losses in the final years would exceed $50 billion for a similar five year period!
40. Evan Ramstad, "Intel Will Introduce New Chip In 1995," *Albuquerque Journal*, January 28, 1994
41. Deborah Baker, "Study: Tax Breaks Can Backfire," *Albuquerque Journal*, September 4, 1993
42. *ibid.*
43. *ibid.*
44. *ibid.*
45. *ibid*
46. "Beauty Of Landing Intel May Be Only Skin Deep," *op. cit.*
47. "Rio Rancho," *New Mexico Business Journal*, May 1992
48. Cord McQueen, "Activists Urge State To Protect Water Supply," *The Observer*, December 29, 1993
49. "Intel Asked Gore For Reassurance," *op. cit*
50. Letter to David L. Shea of Intel, from New Mexico Environment Department, March 16, 1992
51. Citizens Fund, "Air Pollution Facts," *Poisons in Our Neighborhoods: Toxic Pollution in New Mexico*, December 1992
52. Tony Davis, "The Land Of Emissions," *Albuquerque Tribune*, September 23, 1993
53. Lawrence H. Summers, World Bank Inter-Office Memorandum, Subject: GEP, page 5, December 12, 1991
54. Jeff Radford, "Intel Air Quality Violations Will Bring Fines, State Says," *Corrales Comment*, October 23, 1993
55. *ibid.*

56. "Intel Settles Dispute," *Albuquerque, Journal (Update)*, January 29, 1994
57. "Why Tighter Air Standards Are Needed For Intel Plant," *op. cit.*
58. If the reader will allow the conceit, just imagine an "Operation Environment" room on Silicon Mesa . . . a large sign, just like in the Clinton Campaign for President, proclaiming "It's the odor, stupid" reminding Intel's environmental engineers of the "real" issue . . . Well, it's not the odor, stupid!
59. Paul Logan, "Intel Emissions Won't Require Federal Permit," *Albuquerque Journal, Metro Plus,* March 18, 1994
60. Paul Logan, "Intel Comes Under Fire For Present Water Use," *Albuquerque Journal,* April 28, 1994
61. Jeff Radford, "Drawdown From Intel Pumping Will Be Checked In Wells Here," *Corrales Comment,* November 6, 1993
62. Rebecca Smith, "Beauty of Landing Intel May Be Only Skin Deep," *San José Mercury News,* February 22, 1993
63. Jeff Radford, "More Air Pollution From Intel Than At Silicon Valley Plants," *Corrales Comment,* August 7, 1993
64. Paul Logan, "Ambitious Expansion To Create 1,000 Jobs," *Albuquerque Journal,* January 1, 1994
65. Jeff Radford, "The Intel Mentality: Part I, How A Cutthroat Global Battle Affects Corrales," *Corrales Comment,* December 11, 1993
66. Rebecca Smith, "Why Intel Jilted The Golden State," *San José Mercury News,* July 20, 1994
67. Jim Barnett, "Future Of New Plant Hinges On Tax Break, Intel Says," *The Oregonian,* August 15-18, 1994
68. Tony Davis, "Economic, Environmental Fallout Accompany Massive Intel Expansion," *Albuquerque Tribune,* May 4, 1994
69. Paul Logan, "Technical Workers Scarce, Intel Says," *Albuquerque Journal,* September 8, 1994
70. Tony Davis, "Ex-State Official Defends New Job With Intel," *Albuquerque Tribune,* June 23, 1994
71. *ibid.*
72. Jeff Radford, "EPA Sees Intel As Major Pollution Source," *Corrales Comment,* June 11, 1994
73. *ibid.*
74. *ibid.*

75. Seth Brechtel, "Emissions Monitoring Too Costly: Intel," *Albuquerque Journal Metro Plus* June 11, 1994

76. Doug McClellan, "Water-Rights Rule To Change," *Albuquerque Journal*, August 31, 1994

77. Chuck McCutcheon, "Companies Demand Water Guarantees," *Albuquerque Journal*, June 2, 1994

78. Dale Chaney, "Empire Builders," *Albuquerque Journal*, September 4, 1994

79. *ibid.*

80. *ibid.*

81. Kathleen Teltsch, "Small Foundation Flexes Its Muscle," *NY Times Metro* June 13, 1994

82. Henry Goldstein, "The Noyes Fund's Stockholder Revolt: Example for Others," *The Chronicle of Philanthropy*, July 12, 1994

83. Paul Logan, "Power Outages To Cost Intel 'In The Millions'," *Albuquerque Journal*, May 21, 1994

84. "Plant Opens Eatery," *Albuquerque Tribune*, August 15, 1994

85. Steve Brewer, "Intel Will Swap Chip," *Albuquerque Journal*, December 21, 1994

86. *ibid.*

87. Paul Cassel, "I'm Fed Up!," *Computer Scene,* March 1995

Appendix I

SouthWest Organizing Project Community Environmental Bill Of Rights

Right to Clean Industry: We have the right to clean industry; industry that will contribute to the economic development of our communities and that will enhance the environment and beauty of our landscape. We have the right to say "NO" to industries that we feel will be polluters and disrupt our life-styles and traditions. We have the right to choose which industries we feel will benefit our communities most, and have the right to public notice and public hearings to allow us to make these decisions.

Right to Be Safe From Harmful Exposure: We have the right to be safe from harmful exposures imposed on us against our will that would affect our health or disrupt our life-styles. It is our right to have a comfortable life-style, safe from toxic chemicals, other hazardous waste and nuisance. This means having safe water, clean air, and being free of excessive and constant noise from industry.

Right to Prevention: We have the right to participate in the formulation of public policy that prevents toxic pollution from entering our communities. We support technologies that will provide jobs, business opportunities, and conservation of valuable resources. As residents and workers, we have the right to safe equipment and safety measures to prevent our exposure in the community and the workplace.

Right to Know: We have the right to know what toxic chemicals industry, corporate polluters, and government have brought or intend to bring into our communities and workplaces, and how these chemicals will be used. We have the right to know exactly what the methods of prevention and disposal of these chemicals will be.

Right to Participate: We have the right to participate as equals in all negotiations and decisions affecting our lives, children, homes and jobs on the matter of exposure to hazardous chemicals and wastes. We will not allow backroom negotiations and "sweetheart deals." We have the right of access without cost to information and assistance that will make our participation meaningful, and to have our needs and concerns be the major factor in all policy decisions. Government agencies at all levels should be responsive to our needs, provide us with necessary data, and include us in all negotiations with polluters. We have the right to sit at the negotiation table with representatives of the responsible polluters and choose our own representatives. All information should be bilingual because of the multi-ethnic nature of our communities.

Right to Protection and Enforcement: We have the right to participate in the formulation of strong laws controlling toxic wastes and vigorous enforcement of those laws. Government enforcement agencies must enforce all laws and regulations. We also have the right to criminal prosecution of polluters. If a person dies from exposure to chemical poisons in the environment, the responsible party must be prosecuted.

Right to Compensation: We have the right to be compensated for damages to our health, our homes, and our livelihoods. The responsible parties must compensate us for medical costs, effects imposed on our children and our elderly, the loss of our land, jobs, livestock, and the destruction of our homes and environment.

Right to Clean Up: The polluters shall bear the financial burden of clean up. The community shall participate and be an equal partner in developing clean up plans and in monitoring the clean up process. Clean up should take place quickly and the technology chosen will be based on speed and effectiveness and not on low cost. Remedial investigations and feasibility studies will be done in the shortest possible time. We have the right to be insured that our problem is not transferred to other communities. Our homes and our environment shall be restored to the way they were before the polluters chose to pollute them.

Appendix II

Principles Of Environmental Justice

The First National People of Color Environmental Leadership Summit

Adopted October 27, 1991, Washington, D.C.

PREAMBLE

WE, THE PEOPLE OF COLOR, gathered together at this multi-national People of Color Environmental Leadership Summit, to begin to build a national and international movement of all peoples of color to fight the destruction and taking of our lands and communities, do hereby reestablish our spiritual interdependence to the sacredness of our Mother Earth; to respect and celebrate each of our cultures, languages and beliefs about the natural world and our roles in healing ourselves; to insure environmental justice; to promote economic alternatives which would contribute to the development of environmentally safe livelihoods; and, to secure our political, economic and cultural liberation that has been denied for over 500 years of colonization and oppression, resulting in the poisoning of our communities and land and the genocide of our peoples, do affirm and adopt these Principles of Environmental Justice:

1. Environmental justice affirms the sacredness of Mother Earth, ecological unity and the interdependence of all species, and the right to be free from ecological destruction.

2. Environmental justice demands that public policy be based on mutual respect and justice for all peoples, free from any form of discrimination or bias.

3.   Environmental justice mandates the right to ethical, balanced and responsible uses of land and renewable resources in the interest of a sustainable planet for humans and other living things.

4.   Environmental justice calls for universal protection from nuclear testing and the extraction, production and disposal of toxic/hazardous wastes and poisons that threaten the fundamental right to clean air, land, water, and food.

5.   Environmental justice affirms the fundamental right to political, economic, cultural and environmental self-determination of all peoples.

6.   Environmental justice demands the cessation of the production of all toxins, hazardous wastes, and radioactive materials, and that all past and current producers be held strictly accountable to the people for detoxification and the containment at the point of production.

7.   Environmental justice demands the right to participate as equal partners at every level of decision-making including needs assessment, planning, implementation, enforcement and evaluation.

8.   Environmental justice affirms the right of all workers to a safe and healthy work environment, without being forced to choose between an unsafe livelihood and unemployment. It also affirms the right of those who work at home to be free from environmental hazards.

9.   Environmental justice protects the right of victims of environmental injustice to receive full compensation and reparations for damages as well as quality health care.

10. Environmental justice considers governmental acts of environmental injustice a violation of international law, the Universal Declaration On Human Rights, and the United Nations Convention on Genocide.

11. Environmental justice must recognize a special legal and

natural relationship of Native peoples to the U.S. government through treaties, agreements, compacts, and covenants affirming sovereignty and self-determination.

12. Environmental justice affirms the need for urban and rural ecological policies to clean up and rebuild our cities and rural areas in balance with nature, honoring the cultural integrity of all our communities, and providing fair access for all to the full range of resources.

13. Environmental justice calls for the strict enforcement of principles of informed consent, and a halt to the testing of experimental reproductive and medical procedures and vaccinations on people of color.

14. Environmental justice opposes the destructive operations of multinational corporations.

15. Environmental justice opposes military occupation, repression and exploitation of lands, peoples and cultures, and other life forms.

16. Environmental justice calls for the education of present and future generations which emphasizes social and environmental issues, based on our experience and an appreciation of our diverse cultural perspectives.

17. Environmental justice requires that we, as individuals, make personal and consumer choices to consume as little of Mother Earth's resources and to produce as little waste as possible; and make the conscious decision to challenge and re-prioritize our life-styles to insure the health of the natural world for present and future generations.

Appendix III

The Silicon Principles

Prepared by Silicon Valley Toxics Coalition and Campaign for Responsible Technology (Draft—March 1992)

1. Establish a comprehensive toxics use reduction program
    - Phase out the use of CFCs and other chlorinated solvents
    - Phase out all carcinogens, reproductive toxins and neurotoxins
    - Phase out the use of acutely toxic gases
    - Implement in-process acid recycling
    - Develop Toxics Use Reduction plans, materials and waste audits, and mass balance materials accounting

2. Develop health and safety education programs and health monitoring
    - Health and safety training must be sensitive to diversity of workforce
    - Health monitoring must be comprehensive and avail able for public inspection
    - Establish nondiscriminatory transfers for pregnant production workers
    - Earmark 5% of all research and development (R and D) money for environmental, health, and safety programs

3. Work with local communities to establish "Good Neighbor Agreements"
    - Include emergency planning and worst case scenario planning, including transportation planning
    - Provide full disclosure to local communities and regular monitoring, including inspections
    - Establish corporate commitment to hiring, training and promoting local residents

4. Implement a Worker Improvement Program and Economic Impact Statements

- Assure that workers are involved in process design and workplace governance
- Assess environmental, social, and economic impacts of new technologies and new facilities

5. Support national R and D policy directed by civilian (not military) needs
   - Support a change in federal R and D funding from Defense Department to Department of Commerce

6. Establish corporate policies requiring equal standards for subcontractors and suppliers
   - Establish technical assistance and technology transfer to encourage pollution prevention at all stages of production, rather than shift the pollution down the production chain to smaller contractors
   - Hire responsible contractors re: labor and environmental policies

7. Establish corporate standards that are enforced equally domestically and internationally
   - Establish corporate policies that assure full compliance worldwide that meet the strictest standards
   - Require all facilities worldwide to make full disclosure of toxics reporting

8. Establish a life-cycle approach to all manufacturing, from R and D to final disposal
   - Design new products from life-cycle perspective
   - Internalize costs of disposal and guarantee return and safe disposal of all used products

9. Work closely with local communities and workers to ensure full oversight and participation
   - Commit to open partnership with workers and community to assure comprehensive participation

Appendix IV FAB 11 "Ideal" Incentive Matrix
HUMAN RESOURCES

Parameter	"Ideal" Incentive	Deliverable	Savings
Relocation Assistance			
A. Discounted air fare, hotels and car rental	A. 10-20% less than corporate rate	A. 10-20% less than corporate rate	$131,197
		A.1 Airfare Discount letter of intent	37,485
		A.2 Hotel	
		A.3 Car Rental	
B. Discounted moving and storage	B. 50-60% less than standard rate	B. 40%	2,096,000
		B.1 Moving and storage	
C. Discounted mortgage and title fees	C. 1-2% off purchase price	C. 1-2% off purchase price	366,000
		C.1 Mortgage fees	109,800
		C.2 Title fees	
D. New home construction purchase discounts	D. 5-8% off purchase price	D. $1,500 per property	219,000
		D.1 New home discounts	
E. Apartment rent discounts	E. 5-10% off monthly rent and 50% deposits	E. 10-15% discounts; corporate apartments	106,680
		E.1 Corporate apartment discounts	
F. Waive all initial deposits and hookup charges	F. Waive all	F. Waive deposits	86,460
		F.1 Waive utility deposits	
G. Relocation Incentives	G. Rental assistance; counselors; relocation materials; tram passes; bank; tours; welcome booklet; video presentation; road shows	G. Incentive package	62,400
		G.1 Rental assistance	40,000
		G.2 Counselors	42,670
		G.3 Relocation materials	19,800
		G.4 Tram passes	49,500
		G.5 Tours	19,800
		G.6 Welcome baskets	2,000
		G.7 Video presentation	25,000
		G.8 Road shows	5,240
		G.9 Settling in program	242,000
		G.10 Discount program	146,400
		G.11 Realtor package	52,400
		G.12 Banking package	
		H. Use of relocation center	
		I. Realtor referral at origination site	
		Total	**$3,879,832**

See yield model and calculations, Human Resources section, Recruiting Strategy, item K.

Parameter	"Ideal" Incentive	Deliverable
Recruiting Assistance A. Single point of contact B. Funding C. Recruiters D. Maximum flexibility on use of funds	A. Specific person identified B. $500K ('92-'95) C. Recruiters D. Maximum flexibility on use of funds	A. Brandon Harwood/NM Dept. of Labor B. $20K plus "in kind" (letters of commitment included with RFP) C. Not permitted by anti-donation clause of the state constitution D. Matching funds through AED. Copies of pertinent regulations included with RFP.
Screening Assistance A. Single point of contact B. Funding C. Screeners D. Maximum Flexibility	Same as "Recruiting Assistance"	A. New Mexico Department of Labor, Sandoval County Field Office B. Not permitted by anti-donation clause of the state constitution C. Yes. Full service local NM Dept. of Labor office being expanded. D. NM Dept. of Labor will work closely with Intel to customize program.
Training Assistance A. Funding B. Restrictions C. Initial Training D. Retraining E. Education	A. $4.5M over 3 years ('92-'95) B. Restricted only to the project; initial training and retraining C. Included in the above D. Included in the above E. Immediate "in-state" tuition status for employees and dependents	A. $1.7 million B. Unemployed and specific job classes C. Yes D. Yes, plus TV-I's new training facility on site (value--$300,000) E. Currently exists
Labor		A. Lower Labor cost for contract services (NM first in productivity) B. Lower cost of living. More discretionary income per employee. Cost of living - 99.1% of US average C. Experienced construction workforce--quickly mobilized, locally available, existing Intel startup team in place D. Proven ability to install equipment ahead of schedule; CDK and Big J workforce on hand and trained for equipment installation E. Process transferred successfully from tech development to manufacturing F. Proven ability to exceed staffing timelines

Parameter	"Ideal" Incentive	Deliverable
Property Tax	Property tax exemption on all project assets	In place on all existing assets and expansions to existing facility through 2010. Thirty-year exemption on all assets except land expected for new project. Exemption for land through 2010.[1] Value $210,000,0000
Gross Receipts Tax & Compensating Tax	Exemption for purchase of the following: 1. manufacturing equipment 2. R & D equipment 3. consumables[2] 4. building materials 5. WATTS and 1-800 telephone service	1. Exemption available through IRB 2. Exemption available through IRB 3. Available through ITC 4. Available through ITC 5. Yes
Investment Tax Credit	Investment tax credit applied to all taxes based on value of equipment purchased.	Yes, 5 percent of the value of equipment purchased.[3]

NOTES:

[1] If the new Fab is a new project financed by IRB's, state law exempts it from proprety tax for a maximum of 30 years from the date it is leased to the company. A new project would be one separate from the existing plant, not functionally related or subordinate to it. It can be located on the existing project land.

[2] There is no exemption for consumables and building materials, but an investment tax credit is available to apply against gross receipts tax or compensating tax otherwise payable in connection with the purchase of consumables and building materials.

[3] Additional employees must be added. Credit applied to future gross receipts, compensating or withholding tax liabilities or taxpayer may claim refund of gross receipts or compensating tax paid.

TAX IDEAL

Parameter	"Ideal" Incentive	Deliverable
Income/Franchise Tax	1. Low or zero percent rate 2. Investment tax credit (ITC) on equipment 3. Incremental new investment--specific tax credit 4. Double weighted sales tax 5. Factor relief (payroll & property for manufacturing and R & D 6. R & D tax credit 7. 100 percent dividend exclusion 8. Pollution control facility credit 9. Energy conservation credit 10. Moving expense reimbursement exclusion for employees 11. Education expense reimbursement exclusion for employees (not dependent upon federal expiring provision) 12. Child care credit	1. 7.6%[4] 2. Yes, 5 percent (see below) 3. Yes, same as investment tax credit 4. Planned for introduction in the 1993 legislative session 5. Provision exists within current tax statute 6. No 7. No 8. No 9. No 10. No 11. No 12. 30% of cost not to exceed $30,000 per year[5]

NOTES:

[3] Additional employees must be added. Credit applied to future gross receipts, compensating or withholding tax liabilities or taxpayer may claim refund of gross receipts or compensating tax paid.

[4] Lower rates apply for first million dollars of taxable income.

[5] The corporation must pay for care of employees, dependent children, and the child care must occur during the employees normal work hours.

109

Parameter	"Ideal" Incentive	Deliverable
Air Emission	A. Increase TPY under existing permit by 100 TPY B. Fallback No offset No Bact/Laer unless real reduction Facility-wide permit (no individual equipment/process permit) No permit mod. for equipment/process add or changes if within facility emission limit	A. Permit modification submitted at 215 tons additional VOC's and > 100 tons NOx. B. Yes. Exists under current permit.
Water Discharge	A. No storm water permits B. No sewer fees for X (10%) years C. Simple general storm water permit with limited monitoring (scope and frequency)	A. Not applicable. Intel NM EHS made decision to apply for notice of intent. B. Not applicable. See Matrix, Infrastructure, Sanitary Sewer. C. None required
Hazardous Material/Waste	A. Limit waste min. law to 313 list and quantity B. Incorporate federal wastewater treatment exemptions C. Clarify general federal law to exclude out-of-state gens.	A. None apply to NM B. None apply to NM C. None apply to NM
Permit Timetable	A. Guarantee 3 months for permit to construct B. Three additional months to finish all other permits C. Pay for permit fees and processing costs D. Hire/appoint coordinator (+ staff) with government authority to move process	A. Permit expected to be issued December 1992. D. Letter of commitment in original RFP. ED assigned Mr. Vacker to assist process.

Parameter	"Ideal" Incentive	Deliverable
Site	A. A-1 Size; A-2 Cost; no cost > 100 B. Can the site be designated a foreign trade zone? C. Can the existing site be designated an enterprise zone? D. Land use and zoning E. Easements and right-of-ways	A. No cost > .80A B. Approval expected Q193 C. No D. Light Industrial. In place. Construction Permits: 30-45 days, general contractor takes architect's plans to Santa Fe. Released in stages. E. In place
Roadways	A. Off-site improvements needed and cost to Intel B. On-site improvements needed and cost to Intel	A. All offsite improvement met at no cost to Intel B. None required
Utilities Overview	A. Existing utilities site plan B. Proposed utilities site plan	A. Adequate submittals B. Adequate submittals
Electrical	A. Meet the specification for service B. Reliability of service < 1 outage/yr. > 1 min. C. Quality of service[1] < 3 disturbances/yr > 10% voltage change. D. Cost for additional infrastructure E. Schedule for additional infrastructure[1] Non gating < 3 months. F. Estimated rate cost. < $0.06KWH. G. Incentives affected. > 16% reduction, no fees. H. Quality or reliability improvements[1] > 15% improvement	A. Meets and exceeds 150% of specification B. Meets specifications[2] C. Meets specifications. EPRI study underway to characterize quality of service[3] D. No cost to Intel for infrastructure E. Infrastructure installation will not add to standard construction schedule F. Average rate $ .0474; known rate path to year 2000[4] G. None. Intel has the lowest tariff on PNM's system. H. Quality partnership[3]

NOTES:

[1] Parameter not requested in original RFP, has been requested in Revision 1 of the RFP.

[2] System is designed with 100% redundancy. Only one interruption in past five years that exceeded one minute in duration. This interruption occurred during an Intel shutdown.

[3] PNM's system performance exceeds industry average according to 1992 survey of US utilities. Electric Power Research Institute study presently underway to address power quality concerns for wafer manufacturing. Results will be pertinent to all Intel manufacturing facilities.

[4] The applicable tariff for Intel, the Rural Manufacturing Incentive Rate (RMIR) is the lowest tariff on PNM's system. The RMIR provides a known rate path to the year 2000. The rate in 1994 is $ .0427/KWH, escalating to $.0508 at the end of 1999. PNM is willing to explore with Intel the possibility of a rate design based on marginal cost for service after the expiration of the RMIR.

INFRASTRUCTURE

Parameter	"Ideal" Incentive	Deliverable
Water		
A. Meet the specification for service (city water)	A. Meets needs	A. OK. Expansion required to meet 150% peak demand
B. Meet the specification for service (fire water)	B. Meets needs	B. OK.
C. Evaluation of supply system	C. Meets needs	C. OK
D. Cost for additional infrastructure	D. No cost to Intel	D. Paid through rate design per New Mexico state law
E. Schedule for additional infrastructure	E. Non-gating (<8 months)	E. <6 months
F. Estimated rate cost	F. <$0.00/gal	F. .00092/gal.
G. Incentives offered	G. >16% reduction, no fees	G. Declining block rate
H. Quality or reliability improvements	H. >16% improvement	H. OK*
Sanitary Sewer		
A. Meet the specification for service	A. Meets needs	A. OK (2)
B. Evaluation of waste treatment systems	B. Meets needs	B. OK (2)
C. Cost of additional infrastructure	C. No cost to Intel	C. None, in place
D. Schedule for additional infrastructure	D. Non-gating < 8 months	D. In place
E. Estimated rate cost	E. <$0.00048/gal.	E. .000856
F. Incentives offered	F. > 16% reduction, no fees	F. None
G. Quality of reliability improvements	G. > 16% improvement	G. See Environmental
Storm Water		
A. Storm water issues	A. Meets needs	A. Filed notice of Intent for storm water permit 09/30/92
B. Treatment or collection requirements	B. Meets needs	
C. Define needed expansion	C. Meets needs	
D. Cost for additional infrastructure	D. No cost to Intel	
E. Schedule for additional infrastructure	E. Non-gating < 6 months	

*Pressure variations eliminated with latest project upgrades.

INFRASTRUCTURE

Parameter	"Ideal" Incentive	Deliverable
Nitrogen		
A. Meet the specification for service	A. Meets needs	A. On-site generation required
B. Quality of product gas	B. Matches on-site generation quality	B. OK
C. Cost for additional infrastructure	C. No cost to Intel	C. OK
D. Schedule for additional infrastructure	D. Non-gating < 8 months	D. OK
E. Estimated rate cost	E. < $800/MCFM	E. Not given
F. Incentive offered	F. > 16% reduction, no fees	F. -
Natural Gas		
A. Meet the specification for service	A. Meets needs	A. Meets and exceeds 150% of specification
B. Cost for additional infrastructure	B. No cost to Intel	B. No cost to Intel for infrastructure
C. Schedule for additional infrastructure	C. Non-gating < 6 months	C. Construction can be completed 3 months from notification
D. Estimated rate cost	D. < $0.033/therm	D. Total transportation cost $.079/therm
E. Incentives offered	E. > 15% reduction, no fees	E. None
Communications		
A. Meet specification for service	A. Meets needs	A. OK
B. Cost for additional infrastructure	B. No cost to Intel	B. -
C. Schedule for additional infrastructure	C. Non-gating < 9 months	C. 60 days
Zoning and Permitting		
A. Site zoning classification and time to rezone	A. Light Industrial < 2 months	A. OK
B. Current code requirements	B. Current adopted versions	B. OK
C. Visual or line of sight restrictions	C. None	C. -
D. Building height limitations	D. < 86 feet	D. -
E. Additional hookup fees	E. < 4 months	E. None
F. Process for permitting	F. None	F. 1 month
G. Additional fees	G. Minimal	G No fees
H. Site risk (nature's disasters)	H. -	H. OK
Environmental		
A. Noise level restrictions	A. < 70 dB	A. None required by law
B. Protection of flora/fauna	B. None	B. None
C. Environmental impact reports	C. None. No cost to Intel	C. None
D. Wetland issues	D. None	D. None

113

Appendix V

Letter from Bill Garcia to Michael Splinter

STATE OF NEW MEXICO
Economic Development Department

Bruce King
Governor

Joseph M. Montoya Building
1100 St. Francis Drive
Santa Fe, New Mexico 87503
Phone: (505) 827-0300

William E. Garcia
Cabinet Secretary

March 31, 1992

Mr. Michael R. Splinter, Vice President
Intel - Components Manufacturing
4100 Sara Rd.
Rio Rancho, New Mexico 87124

RE: Intel Expansion Project

Dear Mr. Splinter:

I appreciate the consideration Intel is giving New Mexico for the next major expansion, Fab 11. It has been my pleasure to work with Intel in the past and it will certainly be my pleasure to work with Intel in the future as well. Staff members of this Department have been involved in helping build the presentation you have before you today.

I feel the best role for the New Mexico Economic Development Department is to be a tool to be used by companies such as Intel and by communities like Rio Rancho and Albuquerque.. It was our pleasure to push recent legislation to allow the continuation of the Investment Tax Credit, a tool Intel has used extensively in the past and will will have available in the future, should your site selection team recommend Rio Rancho. This business development legislation is unique in the nation and certainly one which has at least helped Intel reach the profitable position the company is in today.

This is but one example of the business climate New Mexico has built in recent years. With this Department's continued effort, I feel we will be able to point out New Mexico's business climate in 10 years as having an even better business climate than today.

Please know that I will do whatever is available to me to ensure that Intel's Rio Rancho plant is as successful tomorrow as I understand it is today. Call upon me at any time if I personally or this Department may be of assistance.

Sincerely,

William E. Garcia

Appendix VI

"Corporations Build on Foundation of Welfare"
Article by José Armas

(A Hispanic Perspective) Reprinted from the *Albuquerque Journal*, December 11, 1994, with author's permission.

Only in America. Supposedly we care about crime and welfare. But, let's check the evidence.

California's Proposition 187 would give license to hunt and persecute undocumented workers whose crime is to do work no one else wants to do. We could have halted this so-called problem yesterday by simply targeting the "criminal" employers and mandating a jail sentence and fines for these outlaws. But are we ready to make suburban mothers, professionals or wealthy businessmen criminals?

We're so concerned with crime that the U.S. Postal Service is about to honor Richard Nixon by putting his face—one of the most notorious criminal faces of our century—on a stamp. We hate crime so much that in this revolutionary election season we've elected as our leaders former convicts or persons under indictment for such crimes as drugs, sexual abuse, fraud and bribery. Witness the election of Marion Barry as mayor of Washington, D.C., and the re-election of Reps. Joseph McDade, R-Pa., Walter Tucker, D-Calif, and Mel Reynolds, D-Ill. Exactly which criminals are we interested in?

We want to end welfare. Is that a question or a statement? U.S. Labor Secretary Robert Reich suggested that while we cut welfare to the poor in this country, we should cut welfare for the rich and powerful corporations as well.

Are we really willing to end welfare for the rich corporations where the most money is spent, where the cuts won't be missed? Noam Chomsky of the Massachusetts Institute of Technology says that our high tech industry is almost totally supported with taxpayers' money. Are we going to cut the billion-dollar welfare bag of goodies by eliminating subsidies, price supports and tax incentives for the rich, giant corporation?

Want an example of one rich corporation that reaps millions in welfare from the government from taxpayers in one of the poorest states in the union?

That's easy: Intel, which has one of the world's biggest expansion projects in progress in Rio Rancho. The welfare Intel receives from taxpayers is outrageous, and no one raises a whimper about it—almost no one.

It would be Christmas year-round for Intel with no complaint if it just weren't for the SouthWest Organizing Project's quest for accountability. Using the Freedom of Information Act, SWOP managed to get a copy of the proposal New Mexico offered Intel to induce them to locate here. It reveals Intel was given $300 million in welfare from taxpayers' pockets for creating 1,000 jobs in its expansion.

'Course, this corporate welfare is called fancy names like "abatement, investment tax credits, corporate tax relief, environmental concessions, ad valorem property tax exemptions, gross receipt and compensating tax exemption on equipment, manufacturing investment tax credits, in-plant training reimbursement for new hires, educational incentives, human resources incentives (such as utility deposit waivers, new home and apartment discounts, discounted mortgage and title fees, etc., etc.)."

By the state's own figures, Intel was offered more than $300 million, which works out to about $300,000 for each job, folks.

There is more.

Intel can hire the majority of its workers from outside New Mexico. The precious water that Intel uses costs much less than what the ordinary homeowner pays. Intel is allowed to pollute our air 35 times more than in California. The 180-acre parcel on which Intel sits will be purchased for $1—and in the meantime, it pays no property taxes. All of which has contributed to New Mexico Intel's profits soaring to an astounding $1.5 billion last year. So, New Mexico taxpayers underwrite the profits for this for-profit company.

Bill Garcia of Intel disputes interpretations of some amounts and adds that "rather than focusing just on Intel, ...certainly there are many other companies who have benefited from (the) same kind of incentives."

And, from the incredible if true department: Intel's welfare for creating 1,000 jobs is $300 million—while this year's entire state welfare budget is $235 million, which serves 393,813 poor adults and children. Who needs the welfare more, rich Intel or the poor whom we give penny ante amounts—or dry milk, rice and beans?

And who do you think is being targeted, corporate welfare or poor people's welfare? Of course, it's the poor. They're easy to attack.

So, we have renewed concern about crime and welfare? Please give me a break from this hypocrisy—at least until after the Christmas season.

Appendix VII

"Good Business Or Public Burden?"
Article by Tony Davis

(first of three sections, printed on July 28, 1994)
Reprinted with permission from the *Albuquerque Tribune*,
(permission does not imply endorsement).

Contrary to state estimates, Intel will get more in tax breaks over the next 30 years than its expansion will generate in tax revenues, our analysis shows.

New Mexico and Sandoval County are giving more in tax breaks to Intel over 30 years than its $2 billion expansion will return in tax revenue.

A *Tribune* analysis shows that the state used questionable number-crunching and outdated and inaccurate figures in defending the tax breaks and in a study saying the expansion will benefit local and state tax coffers.

Some economists say it isn't necessarily bad if Intel's tax breaks exceed revenue, because the plant will plow hundreds of millions of dollars into the private economy.

Others say the state hasn't properly studied the costs and benefits, not just of Intel but of all big private employers whose gateway to New Mexico is paved with tax incentives.

Tax breaks for Intel represent dollars that other taxpayers ultimately will have to pay to build schools, roads, police and fire stations and other public services needed to keep up with growth.

State Senate President Manny Aragón said legislators now question if they "gave the store away" to Intel and other companies and will give future industry requests for tax breaks a more critical eye.

"I don't think people paid too much attention to them in the past. A number of the people moving to Rio Rancho include many young couples who need infrastructure, schools, playgrounds, and we didn't take that into consideration," Aragón said. "You can rest assured we'll be looking at the tax breaks more in the future."

The $1.8 billion to $2 billion Intel expansion will double the plant's payroll from 2,400 jobs in 1993 to 4,800 jobs by 1997. It is the biggest single manufacturing investment in the United States in the 1990s, Intel has said.

For the administration of Governor Bruce King, the Intel expansion has become a big selling point as the governor fights for re-election. King's spokesman, John McKean, argues that Intel's ripple effects on the economy are so great that even without a comprehensive study, all indications are that Intel has been a boon to the state overall. He cited, for instance, other high-tech companies such as Intuit and Sumitomo Sitix that are looking at moving to or expanding in the Albuquerque area.

"What it boils down to is that we have to provide reasonably good-paying jobs for ordinary people in this state." McKean said. "That doesn't mean we should give them at any cost, but in the governor's view the costs are sustainable compared to the benefits we will derive in terms of employment opportunity."

Over 30 years, Intel's Rio Rancho computer chip manufacturing plant will gain about $2 million more in tax breaks from the expansion than it will pay in taxes to the state, Sandoval county and the city of Rio Rancho, The *Tribune* analysis concluded.

The analysis found that the expansion will bring in $287 million in higher tax revenue to state, Rio Rancho and Sandoval County governments over 30 years. The company will get $289 million in tax breaks from the state and county over 30 years.

Either total amounts to about $9.5 million a year, or enough to build two elementary schools for 1,500 students each.

The state Economic Development Department, however, has been predicting that the Intel expansion will net state and local governments $600 million to $700 million.

The differences between the two estimates can be explained like this:

• The state failed to use a common accounting method that accounts for inflation and changes in the value of the dollar to investors over time.

• The state inaccurately claimed that Intel would in 15 years start having to pay property taxes on its expansion.

• The state didn't count most of Intel's tax breaks against the state's and Sandoval County's treasuries. That claim rests on the highly controversial assumption that Intel never would have expanded without the tax breaks—so the state would never have gotten the tax revenue unless the state gave the tax breaks.

• As recently as late April, the state was using year-old estimates of how much money the semiconductor giant would spend on the expansion and how many people it would hire.

But if revenue doesn't outpace tax breaks for companies, individuals have to make up the difference in the form of higher taxes.

"The costs of development and growth are real and must be paid," said Louis Head of the SouthWest Organizing Project, the first group in the state to publicly criticize Intel's tax breaks.

"Higher taxes on the people and/or lower services are a given because of Intel. But the costs of the Intel expansion will not be fully funded by higher taxes because the people won't pay them, they can't afford to."

An economist, however, said that even if New Mexico overhyped Intel's benefits, the expansion probably benefits the state in the long run, considering all the money it will dump into the private economy.

A few million dollars lost in tax revenue over 30 years isn't a big deal, said Tony Popp, an associate New Mexico State University professor.

"I think any government, when they get a big employer, they will try and hype this as much as possible," Popp said. "What New Mexico is saying is that we want these jobs. Even if it breaks relatively even, we've got the jobs and it gives Intel a bigger incentive to keep things here."

Intel and state officials defend the tax incentives as essential to broaden the state's economic base, which today leans heavily on government spending.

"Big employers such as Intel today compete globally, and tax incentives are investments in job creation," says Jonathan Krebs, Deputy Economic Development Secretary.

Since Intel chose Rio Rancho in 1993, Texas and California have approved new tax breaks and subsidy programs similar to New Mexico's, Krebs noted. He likens the bidding war for jobs to a game of chicken in which two states are plunging head-on at each other and "whoever flinches (by not offering enough tax breaks) loses."

The state economic development agency didn't give Intel

the tax breaks. The Legislature approved most of them over the past two decades. Indeed, it was Sandoval county that granted Intel the biggest break, the 1993 approval of a $2 billion industrial revenue bond.

Now, however, the Economic Development Department is in the thick of the Intel controversy because it helped line up the Intel expansion and it has defended the tax breaks against critics. Its former secretary, Bill Garcia, came under criticism last month when he announced he was leaving the state to take a job managing Intel's government and public relations.

In analyzing the tax breaks and revenues for Intel, the *Tribune* also found that:

• Even using raw dollars, the Intel expansion will cost the state more in tax breaks than it gains in revenue for its first 15 years.

And critics question whether Intel still will be around or running at today's scale in 15 years, let alone 30 years, given the volatile nature of high-tech and electronics industries.

As ammunition, they point to the story of Siemens-Stromberg-Carlson, a telephone equipment manufacturing company. It fled its Albuquerque plant for Florida and laid off 300 workers in 1993. That came barely a year after Albuquerque City Hall granted an $11 million bond for Siemens that provided tax breaks to retool and repair the Northeast Heights factory.

Critics also have questioned whether such tax breaks should be given at all, to any industry.

"What will be the structure of these industries in 15 years?" said Anna Lamberson, a Land Commissioner's office economist. "Are they going to be there? Somebody needs to address these questions."

Intel says it's "nonsense" to think that it will pull out anytime soon.

"I can't guarantee they'll be here in 40 or 50 years, but look at our track record," Intel spokesman Richard Draper said. "During the volatile 1980s, we still kept gearing up here. We retrofit. We put in new equipment. We don't just shut down."

• The SouthWest Organizing Project, a six-person-nonprofit activist group with an annual budget of $170,000, was closer to the mark in a recent estimate of Intel's tax breaks than was King's Economic Development Department, whose annual budget is $8.76 million. In late April, the organizing project estimated that Intel would get $250 million in raw dollars in tax breaks over five years. At that time, the state was predicting $198 million in tax incentives over 18 years.

The organizing project made errors, but not as big as the state made. It mistakenly assumed that Intel would pay the property tax rate for Rio Rancho. That rate is higher than in unincorporated Sandoval County, where Intel's plant sits in a government no-man's land bordering Rio Rancho and Corrales.

Garcia used outdated estimates of Intel's tax breaks in an April interview to counter the organizing project's higher estimates.

The older estimates that Garcia used assumed that Intel would hire fewer than half as many employees—1,000 instead of 2,400—and spend only a little more than half as much as it will actually spend on the expansion. Using higher estimates for the expansion would have led to higher estimates for the tax breaks.

When Garcia gave the interview, his staff's files contained information that showed his older estimate was outdated. The files showed that Intel's jobs and tax breaks would be higher than his staff had predicted a year earlier.

As late as mid-May, the state also was basing its estimates of Intel's property tax breaks on 1991 Sandoval County property tax rates, now three years out of date.

As he was preparing to leave his state job for Intel, Garcia said he had relied on his staff to keep him abreast of the numbers and that they hadn't told him that the numbers should change.

However, Garcia said he accepted "full responsibility" for giving out the outdated numbers.

• The state mistakenly assumed that starting in 2010, Intel would lose half of its annual property tax break.

That's wrong. Sandoval County freed the company from paying any property taxes to Sandoval County on land, equipment or buildings from the expansion until 2023, says County Manager Debbie Hays.

Krebs said that he got that information from Intel in response to his questions.

"They misunderstood what I was trying to get (at)," Krebs said.

State agency didn't do complete cost-benefit study

(second section of Tony Davis article)

Albuquerque City Hall does it regularly. Once in a while, the booming suburb of Rio Rancho does it.

But the state didn't look at some of the biggest public costs or benefits of the Intel Corp.'s $2 billion expansion project when it concluded that the project would be a big boon to state and Sandoval County tax coffers.

Rio Rancho has done it once, and Albuquerque City Hall has done it several times for big new manufacturers or real estate developments. Sandoval County, where Intel's plant sits outside the Rio Rancho city limits, hasn't done such studies.

As Albuquerque purse strings tightened over the past two decades, city planners increasingly examined how much it costs in roads, water and sewer lines to serve a new development.

"We look at our infrastructure costs to the penny," said Shirley Wozniak, a City Hall senior planner and economist. "There's no reason the state can't do that."

But the state Economic Development Department has not looked at:

- The costs of building roads and laying water and sewer lines to the Intel plant and its employees' homes.
- The dollar value of the additional water, nearly 3 million gallons daily, that Intel will pump from the ground for its expansion.
- The costs of the expansion's increased air emissions, although the company is trying to reduce the pollution by installing incinerator-like-devices.
- The extra tax revenue from "spin-off" jobs that Intel generates by attracting other companies.
- The corporate income tax revenue and corporate income tax breaks stemming from the expansion, because those figures aren't available.
- The effects of Intel's expansion on the private economy.
- The social benefits of more employment.

Recently, a Washington, D.C., economist predicted in a study for Intel that its expansion will stimulate other New Mexico employers to create 6,300 more jobs by 1997. All industries in the state would increase their production by $1.18 billion because of Intel, and citizens would earn an extra $194 million, said Robert Johnston, an economist for J.W. Wilson & Associates Inc.

But the University of New Mexico's Bureau of Business and Economic Research contends that Johnston, who used U.S. government computer models, is too optimistic. Bureau officials, who use their own computer model, contend that one to 1 1/2 "spin-off" jobs per new Intel job is a more realistic estimate.

"There is too little information known now to be able to tell if Intel is a positive or negative to the state, overall," said Anna Lamberson, a state Land Commissioner's office economist.

The state Economic Development Department does limited cost-benefit analysis "whenever we're asked to do it," said Jonathan Krebs, the department's deputy secretary. It would be difficult to come up with a computer model that could work anywhere, Krebs said.

He contends it's not fair to load the costs of growth onto one company, because society has decided it wants growth and everyone wants better roads and schools.

"Would Rio Rancho have decided they did not want to start their own school district if Intel had not come? Would the citizens of Sandoval County have been satisfied to leave Highway 528 as a two-lane?" Krebs said.

But when Albuquerque City Hall put several real estate developments up to its cost-benefit test, not all of them passed.

"Not to know the economic effects is to have a piece of the puzzle conspicuously missing. It isn't a deciding factor, but it is one more piece that decision makers should have," city planner Wozniak said.

Rio Rancho also looked at the costs of services in September 1992 when it issued an industrial revenue bond to a new Aeroparts Manufacturing and Repair Inc. manufacturing plant that makes aircraft sheet metal components. The bond, similar to the one Sandoval County issued Intel, frees the plant from paying property taxes for 30 years.

"Obviously, a study like that would have to look at infrastructure. They would have to look at everything we would lose or we would gain, and you come to a bottom line either it is a good deal or not a good deal," City Administrator Hal Donovan said.

Lamberson, Brian McDonald of UNM's Bureau of Business

and Economic Research, Tony Popp of New Mexico State University and several other economists agree that the state should look at costs of serving a new employer and it's workers.

"The main reason that boom towns in the West have trouble keeping up with growth is that your property taxes generate a big flow of income over time, but you need to make substantial capital investments up front," said Mike McKee a UNM economics professor. "If you go into a bust cycle, you have the commitments you've made for streets and roads, but you don't have the taxes to pay for it."

Trib analysis uses accepted accounting method

(third section of Tony Davis article)

Why are the *Tribune's* and the state's estimates of Intel's tax breaks and tax revenue so different?

The differences stem from three factors:

• The state didn't use a standard accounting method of discounting the value of the tax breaks and the tax revenue that the Intel plant will generate over 30-years. The *Tribune* did.

• The state didn't count most of Intel's tax breaks as costs to the state. The *Tribune* did.

• The state mistakenly assumed that the semiconductor giant would pay $100 million in property taxes that it won't have to pay.

Many economists and accountants contend that the *Tribune's* tax number-crunching using what's called "present value discounting," is more valid than the state method using expected raw dollar totals over 30 years.

The discounting totals the numbers in current, 1994 dollars. This technique is supposed to work well for investors, for whom a dollar today is more valuable than a dollar they will get in five, 10 or 30 years.

The *Tribune* had accountants in Albuquerque and Santa Fe do the discounting. The accountants, who requested anonymity, came within a few thousand dollars of each other. The *Tribune* verified the numbers.

In raw dollars, the money that Intel is likely to send into state and local treasuries over 30 years will exceed the plant's tax breaks by nearly $100 million.

But in discounted dollars, the tax breaks will outpace revenue by $1 million to $2 million over 30 years. The reason is that far more of the tax breaks occur in the expansion's early years and more of the revenue will go to the state in later years. The discounting weights the value of current dollars more heavily.

The *Tribune* used a 7 percent discount factor, a little less than the interest rate someone could get by buying 30-year government treasury bonds.

Jonathan Krebs, the state's deputy Economic Development secretary, said in a letter to the *Tribune* that he didn't discount his study's costs and benefits over 30 years because "I don't know how."

"I don't frankly understand present value calculations," Krebs said.

But he contends that only two incentives, worth $57 million in raw dollars and about $50.7 million in discounted dollars, should be counted as costs to the state.

If Intel hadn't expanded, the state would have saved $55 million in investment tax credits. If that tax credit didn't exist and even if the expansion hadn't happened, Intel would have had to pay those taxes in gross receipts taxes and in withholding of employee's income taxes.

The state also would have saved $2 million it now plans to spend in state subsidies to train Intel employees, Krebs said.

But the state shouldn't count as costs, two other big Intel tax incentives, Krebs contends. One gives Intel a 30-year break on its property taxes to Sandoval County. Another gives Intel a break on sales taxes for the machinery that Intel buys to make its computer chips.

The state wouldn't have reaped this money if it hadn't given Intel the tax breaks, Krebs said, because Intel wouldn't have developed its land or bought this equipment unless it had expanded. Then, the company wouldn't have had to pay those taxes.

University of New Mexico economists Brian McDonald and Mike McKee and New Mexico State University economist Tony Popp, however, said the state should count all the incentives as costs.

"If the state had said no to the incentives, Intel might have gone back and sharpened their pencils and said that even without tax incentives, it might have made sense to go to New Mexico," McDonald said. "But we'll never know."

But without the tax breaks, New Mexico's chances of landing the Intel expansion could have been in jeopardy, company spokeswoman Barbara Brazil said.

The incentives, the presence of Rio Rancho's existing Intel plant and a highly qualified work force and construction crew weighed heavily for New Mexico against five competing states, Brazil said. Those states were Arizona, California, Oregon, Texas and Utah.

INTEL TAX BREAKS
- **Property Tax Breaks:** $441 million over 30 years from Sandoval County. Source: 1993 $2 billion industrial revenue bond.
- **Sales Tax Breaks:** $70 million over five years from New Mexico. Source: 1993 industrial revenue bond.
- **Investment Tax Credit:** $55 million over three years from New Mexico. Source: state law.
- **Employee Training Funds:** $2 million over two years from New Mexico. Source: state program.
- **Corporate Income Tax Breaks:** Unknown. Will take effect in 1995. Source: state law.
- **Total Tax Breaks:** Not counting corporate income tax breaks, $568 million over 30 years in raw dollars and $289 million over 30 years in dollars discounted to reflect the declining value of money over time.

INTEL TAX REVENUE
- **Income, Property And Other Taxes From Permanent Employees:** $254 million over 30 years.
- **Gross Receipts Tax From Purchase Of Construction Materials And Labor:** $42 million over three years.
- **Income, Property And Other Taxes From Construction Jobs:** $10,839,543 over three years.
- **Gross Receipts Taxes From Intel Purchases Of Goods And Services:** $216 million over 30 years.
- **Gross Receipts Taxes From Gas, Electric And Phone Bills:** $51 million over 30 years.
- **Payments In Lieu Of Property Taxes:** $5,250,000 over 17 years.
- **Income And Property Taxes From Indirect Retail Jobs That Intel Generates:** $79,040,832 over 25 years.
- **Corporate Income Tax:** Unknown over 30 years. $20 million in 1994. $11 million increase over 1993, amount of increase due to Intel expansion unknown.
- **Total Tax Revenues:** Not counting corporate income taxes, Intel will generate $659 million in raw dollars and $287 million in discounted dollars over 30 years.

Appendix VIII

"Small Foundation Flexes Its Muscle"
Article by Kathleen Teltsch

Reprinted with permission from the *New York Times*, Metro Section, June 13, 1994

They are curiously matched adversaries.

One is Intel Corporation, which is building the world's largest computer-chip plant at Rio Rancho, N.M. The other is the Jessie Smith Noyes Foundation, a small family philanthropy based in Manhattan, which owns 3,000 shares of Intel stock.

The two came face to face at Intel's annual stockholders' meeting in an encounter that in the foundation world is considered a bold use of financial muscle.

Stephen Viederman, the foundation's president, exercised his rights as a shareholder to question "our company" on its rosy forecasts of benefits from the $1.8 billion plant expansion, which Intel said would create 2,400 manufacturing jobs and bring enormous economic benefits to New Mexico.

A number of major foundations say they are watching the Noyes initiative to see how effective it is.

"We're following it with interest," said Peter C. Goldmark, the president of the Rockefeller Foundation. "I've always had grave doubts that passive ownership of stock is an effective lever for change, but maybe Steve can prove otherwise."

Raising Tough Questions

Foundations are minor figures in the investment field: collectively they own only 2.2 percent of all publicly traded stock.

And until now, foundations have protested quietly, if at all. Trying to influence corporations to be more socially responsible, some have divested themselves of stock or joined shareholders' resolutions that take issue with companies on political or social questions.

"The Noyes Foundation's appearance at the Intel stockholders' meeting is the first time a foundation executive ever has taken such a step and raised tough questions with corporate management," said Timothy H. Smith, the director of the Interfaith Center on Corporate Responsibility in Manhattan. "Noyes is on the cutting edge of a new trend."

The 23-year-old Interfaith Center has sponsored thousands of shareholder resolutions, often drawing support from church groups, universities, trade unions, pension funds and, over the years, about 50 of the nation's 35,000 foundations.

The role of visible and vocal campaigner is a new one for Noyes. With $60 million in assets, it ranks 351st in foundation wealth. Noyes has a history of involvement in environmental issues. It has supported traditional agriculture on lands of the Zuni pueblo in New Mexico and worked to protect the Brazilian rain forest.

The foundation was created in 1947 by Charles F. Noyes, a prosperous real-estate broker. (His most famous deal was the 1951 sale of the Empire State Building, considered a white elephant at the time.) The foundation, with headquarters at 16 East 34th Street, was named for his wife, a volunteer who worked to promote racial and religious rights.

The foundation was drawn into the Intel controversy because it is a financial supporter of the SouthWest Organizing Project, an advocacy group that has picketed at Intel plants in California and New Mexico.

Before the Intel stockholders' meeting in May, the Organizing Project issued a 60-page report, "Intel Inside New Mexico," describing the company as a modern Trojan horse, a gleaming chrome exterior concealing a chemical factory.

The advocacy group charged that New Mexico's state officials, in their eagerness to attract the giant semiconductor for manufacturer to Rio Rancho, had offered excessive inducements including $250 million in tax breaks, and loosened environmental regulations to permit the company to use 90 chemicals, some never tested.

In an interview, Mr. Viederman called the Intel agreement with the state "an important case study of contemporary smoke-stack-chasing, with the environment and the community as victims."

At the shareholders' meeting, Craig R. Barrett, Intel's chief operating officer, characterized the Organizing Project as a "small, vocal minority of people." He said Intel was a good corporate citizen, was open about its environmental practices, had reduced chemical use and listed the chemicals it used.*

An Intel spokesman, Howard High, said the company refused Mr. Viederman's request that it open talks with the Organizing Project and charged that leaders of the advocacy group had come to the meeting, some from outside the state, only to stage a protest for television cameras. In a telephone interview, Mr. High said Intel would try to answer the concerns expressed by Mr. Viederman "as we do any shareholder."

Mr. Viederman said the Noyes foundation would continue to exert pressure until it achieved some satisfaction.

*Publisher's Note: Ironically, one week later, the SouthWest Organizing Project was notified that the organization had received the City of Albuquerque Human Rights Board 1994 *Bridge Award* "in recognition of its long standing dedication, advocacy, and tireless efforts in working to ensure equity for all, working in a spirit of intergroup cooperation and across racial, ethnic, and cultural lines; its accomplishments and contributions have touched many people."

Glossary

acequia: a dirt ditch used for irrigation, usually part of a larger irrigation network connecting fields to a nearby river; acequias are often communally managed and have been used in New Mexico for over three hundred years.

acetone: a volatile, fragrant, flammable liquid used chiefly as a solvent and in organic synthesis; acetone may be inhaled or absorbed through the skin and can irritate the eyes, nose, throat, and mucous membranes; it can also damage the nervous system.

acid neutralization: addition of chemicals to change acidity of liquids; strong chemicals combined with acids to produce salts which dissolve in water.

acid recycling: processing acids used in manufacturing so that they can be reused.

acutely toxic gas: a gas that is immediately harmful or fatal; one brief exposure can damage health.

adrenal gland failure: malfunction of either of a pair of adrenal glands, which are located next to the kidneys; adrenal glands produce adrenaline, hormones which control metabolic functions, and sex hormones; failure results in delayed puberty, sexual dysfunction, etc.

air basin: a geographical formation which gathers and contains an air current system for an extended area of land.

air emissions: substances such as chemicals or heavy metals which are released into the air.

air pollutant: any airborne contaminant, usually in the form of gas or smoke, which may poison land, surface and groundwater, plants, and living beings.

air scrubbers: devices in smokestacks or exhaust systems that remove most, but not all, particles before they are released into the atmosphere.

Albuquerque Box syndrome: airflow pattern over Albuquerque; upper air currents move south, lower currents move north.

ammonia: a pungent, colorless, gaseous alkaline compound of nitrogen and hydrogen most commonly found in the form of a water solution; contact with ammonia may result in irritation of the respiratory system and mucous membranes, burning and blistering of the skin, headaches, nausea, and vomiting.

AMREP: (American Real Estate and Petroleum) Original developer of Rio Rancho.

aquifer: a layer of permeable rock, sand, or gravel which holds large quantities of water.

arroyo: a channel cut into the earth by erosion, also called a wash.

arsenic: a solid poisonous element that is commonly metallic steel gray, crystalline, and brittle; also a poisonous compound of arsenic and oxygen often used as an insecticide or weed killer; arsenic is stored in the body after exposure and can result in conjunctivitis, visual disturbances, irritation of mucous membranes, skin problems, nerve damage, cancer, kidney and liver disorders, and intestinal malfunctions.

arsine: a colorless, flammable, extremely poisonous gas with an odor like garlic; arsine may be produced by action of acids on metals containing arsenic as an impurity, for example, when cleaning metal tanks which have contained acid; it is also used in the manufacture of semiconductors; it is used as a dopant in wafer fabrication. If inhaled, arsine may cause headaches, dizziness, comas, shortness of breath, heart, liver, blood, and kidney damage.

automation: in a workplace, the introduction of machines which perform jobs previously held by human workers.

Best Available Control Technology (BACT): equipment and

methods that remove or eliminate as many pollutants as possible.

calcium hydroxide: a white crystalline compound used in making alkalies, bleaching powder, etc.; also known as slaked lime; used to neutralize acids.

carcinogen: any substance which causes cancer.

carpal tunnel syndrome: Swelling of tendons in the wrist which traps and pinches the nerve. This can cause loss of touch in surface areas of the hand, numbness, pain, and tingling in the thumb and fingers, and loss of strength in the fingers. Usually associated with workers that regularly use computer keyboards over a lengthy period of time, it can affect anyone whose job requires the same hand movement over and over.

cellosolve acetate: a toxic glycol ether used as a solvent and in making computer chips.

central nervous system: the brain and spinal cord, which receive sensory impulses from nerves and which send out motor impulses.

central processing unit (CPU): The part of a computer that does the math and logic calculations. Current preferred term is PROCESSOR.

cervical precancerous tissue: abnormal cells occurring on the cervix, which connects the vagina to the uterus. Will become cancer cells if untreated.

chemical vapor deposition chambers: chamber where thin layers of material are placed on silicon wafers in chip making.

Chicano(a): a self-descriptive term used by people of the Southwestern U.S. who are of mixed ancestry indicating Native American/Mexican Indian and Spanish heritage.

chip: common term for integrated circuits used in computers; a piece of semiconductor material with electronic circuitry etched into it; Processor chips and memory chips are most common, although any electronic circuit can be put on a chip (amplifier, receiver, etc.)

chlorinated solvents: compounds of various hydrocarbons and chlorine; used in wafer and circuit board fabrication and cleaning; effects of exposure include cancer of the liver, lungs, skin, and blood, and disorders of the liver, kidney, heart, and central nervous system.

chlorine: one of several highly reactive elements known as halogens. Chlorine is usually found as a heavy greenish yellow gas of pungent odor and is used especially as a bleach, oxidizing agent, and disinfectant in water purification; it is used in the fabrication of computer chip wafers and is a by-product in the manufacture of light emitting diodes; high levels of chlorine in water has been found to cause bladder cancer, while other types of exposure can cause severe irritation of the skin, eyes, and mucous membranes, severe lung problems, headaches, nausea, and vomiting; chlorine may combine with moisture in a reactive action to form hydrochloric acid.

chlorofluorocarbons (CFC): chemical compounds made from chlorine, fluorine, and carbon; used as lubricants, refrigerants, and as propellants in aerosol sprays. When released into the atmosphere, they rise and damage the ozone layer. The chlorine, fluorine, and oxygen can then combine with water in the atmosphere to produce acid rain. Because of the detrimental effects, an international agreement was made forbidding their use as propellants. It will phase out their production and use by 1996.

chronic organic brain dysfunction: any damage to the brain that is permanent and irreversible.

Clean Air Act: federal law passed in 1970 with later amendments. Regulates pollutants in the air. Recent reauthorization of the Act has expanded the regulation of toxic air emissions.

Clean Water Act: common name for the Water Pollution Control Act of 1972 and its amendments. Sets limits and penalties for dumping pollutants into surface and groundwater.

continuous emissions monitoring: checking amounts of pollutants and other materials in air and water discharges all of the time instead of random, infrequent testing.

deionized water: highly purified water.

demographics: relating to the statistical study of human populations, especially in reference to size, density, ethnic makeup of populations, etc.

detoxification: the process of removing poisons or toxins.

diesel fuel oil #2: commonly used as fuel for emergency generators; usually stored in large underground tanks.

diborane: a potentially lethal chemical used to make silicon wafers; exposure can cause coughing, bleeding in the lungs, headache, dizziness, fatigue, muscle spasms, and nausea.

1,1 dichloroethylene (DCE): a chlorinated solvent used in manufacturing, affects the kidneys and liver.

diffusion furnaces: furnaces which use dangerous gasses at high temperature to implant a chemical into semiconductor material (usually silicon) to create electrical circuits on each chip in a wafer.

dopant: a deliberate chemical impurity introduced to give chips their useful properties.

dynamic random-access memory (DRAM): a type of memory chip. Stored data is lost when the power is turned off.

ecological systems: interdependent relationship between plants, animals, and the environment they live in.

economic development (strategy): generally refers to state and local government efforts to stimulate job creation, reduce unemployment and increase incomes and tax revenues. The most generally used tools are a variety of tax incentives and abatements. Such policies often have built-in contradictions–tax incentives can shift tax burdens to other businesses and citizens. Companies often place a higher premium on services, infrastructure and environment, all of which may require increased payments or taxes when looking at a place to relocate or expand.

economic justice: a conviction that economic policies must result in benefits which are distributed equally across income and racial lines and that jobs created by state and local tax incentives must go to local citizens and taxpayers and that the health, natural resources, and the culture of the community must be protected.

Electronics Industry Good Neighbor Campaign (EIGNC): A joint project between the Southwest Network for Environmental and Economic Justice and the Campaign for Responsible Technology, to promote community accountability of high-tech industries.

encephalopathy: a disease of the brain, especially one involving alterations of brain structure.

environment: where we live, where we work, where we play.

Environmental Impact Report: California equivalent of federal Environmental Assessments or Environmental Impact Statements; evaluation of the effects from development or construction; required for various projects, especially those that receive federal funding.

environmental justice: equal and fair access to a healthy environment; equal enforcement of environmental regulations; and a movement to protect communities of color from environmental hazards (see economic justice).

environmental permitting process: conducting studies, preparing reports, establishing safeguards, etc. to receive permits prior to activities that have the potential to damage the environment; may occur at the federal, state, and/or local level.

Environmental Protection Agency (EPA): an independent federal agency; responsible for setting standards and limits, issuing permits, monitoring and enforcement, supporting research, and assessing the consequences of activities or projects that cause pollution or damage the environment.

environmental racism: racial discrimination in environmental policy-making and the enforcement of regulations and laws, the deliberate targeting of people of color communities for toxic waste facilities, the official sanctioning of life threatening presence of poisons and pollutants in our communities, and the history of excluding people of color from the leadership of the environmental movement.

environmental safeguard: equipment or procedures that provide protection from pollution or the misuse and waste of natural resources.

2-ethoxylethyl acetate: solvent; see cellosolve acetate.

ethylene glycol: a solvent used in cleaning, wafer fabrication, and computer assembly: exposure can result in brain damage, respiratory failure leading to heart failure, and abnormalities in bone marrow.

ethyl-3-epoxy-propionate (EEP): airborne pollutant.

European protectionism: see Protectionism.

etchant: any of a number of corrosive chemicals which strip away a layer on the surface of a wafer, embedding a pattern for electrical circuitry onto its surface.

extractive industries: involve the removal of natural resources; mining, logging, etc.

FAB (fabrication plant): any wafer fabrication plant; FAB is Intel's term for their factories that produce computer chips from silicon wafers.

FAB 11: Intel's expansion FAB in Rio Rancho, New Mexico.

fiefdom: something over which one has exclusive rights or exercises control.

floating point unit (FLP): specialized circuit for calculations using 1 or more decimal points. Can be on single chip or part of processor chip. Also called MATH CO-PROCESSOR.

Foreign Trade Zone (FTZ): specially designated area with few or no tariffs or import restrictions. Raw materials or component parts may be brought in and processed or assembled. Completed products are shipped out for further work or to sell. Taxes are not paid until finished product is sold.

Freon: a trademark name used for various nonflammable gaseous and liquid fluorinated hydrocarbons used as refrigerants and propellants for aerosols. See CFC.

General Agreement on Tariffs and Trade (GATT): Formed in 1946 to promote expanded trade through reduced tariffs and coordinated trade policies. GATT is a series of agreements negotiated through its own international body. Eight rounds of negotiations have been held since 1948. The Uruguay Round was begun in 1987 and completed in early 1994. It is now being ratified by each member country before it takes effect. Beginning in the 1970s, negotiations have also tried to reduce nontariff trade barriers such as government procurement policies, import licensing, and subsidies.

gentrification: the process whereby a neighborhood or commercial district is infiltrated by higher income residents and businesses, usually rendering the neighborhood's rents, property taxes, and cost of living unaffordable for those who have lived or worked in that area previously.

glycol ether: a type of solvent including compounds such as cellosolve, methyl cellosolve, and butyl cellosolve; exposure can cause weakness, anemia, headache, tremors, brain disease, and liver and kidney damage.

Good Neighbor Agreement: Formal agreement between companies and the communities where they are located, promising ethical business practices and allowing co-signers to participate in decisions that will affect them.

groundwater contamination: poisoning of water which lies below ground level. Anything that reduces the quality of water in aquifers.

high tech industry: usually refers to firms involved with computers, electronics, lasers, advanced technology, or research and development as well as the software industry. Often contrasted with low tech industries such as steel mills, coal mines, and auto manufacturing. The work environment is less dirty and noisy and not as noticeable from the outside. Often claimed to be less dangerous and polluting, this is not true. The dangers are not obvious and damages may not be visible for many years. High tech manufacturing industries commonly use materials that are highly toxic and the long-term health and environmental effects of many of them are unknown. The majority of jobs created by high tech industries are low-paid, less-skilled positions that can include regular contact with toxic substances or manual, repetitive assembly-line work.

hydrochloric acid: a water solution of hydrogen chloride which is a strong corrosive acid, used in many stages of computer fabrication; exposure can cause skin, nose, and mucous membrane irritation, nasal erosion and ulcers, coughing, laryngitis, and bronchitis.

hydrofluoric acid: a water solution of hydrogen fluoride commonly used in computer fabrication; even small amounts splashed on skin cause debilitating burns.

hydrogen fluoride: airborne pollutant that can combine with moisture in the air to form hydrofluoric acid.

hydrogen peroxide: a compound of hydrogen and oxygen used as an oxidizing and bleaching agent, an antiseptic, and a propellant. It breaks down into water and oxygen during use.

hypothyroidism: deficient activity of the thyroid gland, resulting in a lower metabolic rate and general loss of energy. In children it causes growth problems.

indigenous people: the original residents of an area.

Industrial Revenue Bonds (IRBs): used to finance industrial development. City, county, or state governments sell bonds to investors. The money goes to build facilities the government will rent to private firms. The rent revenue is used to repay the investors. If there is not enough revenue, the government may be responsible for repaying the investors. Ideally, tax revenues are not supposed to repay the debt. Since a city or state owns them, the facilities have certain tax exemptions. In New Mexico, Intel and other firms have used IRBs solely for tax avoidance.

in-plant incineration: burning industrial waste where it was produced. Reduces pollution by creating smaller volume of concentrated, less-toxic waste, however the burning process produces additional pollution.

integrated circuit (IC): a single semiconductor device containing numerous electrical functions combined to form a complete circuit in a very small space. Most ICs are an inch square or smaller and a quarter inch thick.

Intel "Ideal" Incentive Matrix: a refined and comprehensive list of the most perfect set of tax incentives, infrastructure, resources, labor concessions, which Intel presents to prospective communities/governments in their effort to maximize their profitability–see jobmail

Isleta Water Quality Standards: under provisions of the Clean Water Act, the Pueblo set standards for surface- and groundwater quality that have been approved by the Environmental Protection Agency. These standards are the same as the federal Drinking Water standards, but are higher than the current standards for discharges from wastewater treatment facilities. Albuquerque and other communities upstream from Isleta will have to improve their sewage treatment plants or face penalties for exceeding pollution limits in water that is returned to the Rio Grande.

irreversible organic mental syndrome: permanent changes that affect normal brain functions.

isopropyl alcohol: a solvent alcohol used in many stages of building computers.

jobmail: explicit attempts by a company to obtain the maximum package of tax incentives and subsidies and infrastructure from state and local governments and the local labor force in order to induce the company to locate a facility or to retain a facility within a community.

liquid arsenic: less toxic alternative to arsine gas used in chip manufacturing.

Local Emergency Planning Committees (LEPCs): groups responsible for preparing for disasters. They are required by Federal law.

lupus: any of several diseases characterized by skin lesions

magnetic bubble: a form of computer memory that was developed but was not commercially successful.

mainframe: term for very large computers.

mass balance materials accounting: making sure that everything used in manufacturing produces a finished product and the least amount of waste; the inputs should equal the output. The main goal is to reduce waste.

microelectronics: a branch of electronics that deals with the miniaturization of electronic circuits and components.

microprocessor: a computer on a single semiconductor chip, or an integrated circuit capable of functioning as a small computer. The heart of a PC. The first successful chips were the 8086 and the 8088, introduced in the 1970s. Improvements produced the 286, rapidly followed by the 386, and in less than a year the 486. The fifth generation chip, Intel's Pentium, is thousands of times more powerful than the first chips and has almost made the 386 and 486 versions obsolete.

minicomputer: computers that are more powerful than PCs but not as large as mainframes. Often used in medium sized companies, they are made by DEC, Burroughs, IBM, and others.

MRI: magnetic resonance imaging. A non-invasive, non-radioactive method of taking pictures of the inside of objects, including the human body.

National Priorities List: More than 1200 properties in the US determined by the EPA to pose the greatest threat to human health and the environment. Known informally as "Superfund sites."

n-butyl acetate: a mildly irritating ester solvent used in wafer fabrication and computer assembly.

neurotoxin: a poisonous substance that is harmful to nerve and brain cells.

New Mexico Investment Tax Credit (NMITC): "unique" economic development tax incentive which allows a company to offset 5% of the value of qualifying equipment from tax obligations to include employee withholding liability (see tax abatement).

New Mexico State Emergency Response Commission (SERC): state agency responsible for LEPCs; see Local Emergency Planing Committees.

nitrogen: a colorless, tasteless, odorless, gaseous element that constitutes 78% of the Earth's atmosphere by volume and occurs as a constituent of all living tissues in combined form; used in pure form in wafer fabrication, inhalation may cause death by asphyxiation.

n-methyl-2-pyrrolidone: a solvent used in cleaning and wafer fabrication.

nominal ownership of real estate: land technically owned but actually used and controlled by someone else.

Non-Precursor Organics (NPOCs): organic compounds that do not break down after use and disposal, may be hazardous.

North American Free Trade Agreement (NAFTA): treaty between Canada, United States, and Mexico that reduces or eliminates tariffs on products shipped between the three countries.

ombudsman: a government official or other person appointed to receive and investigate complaints made by individuals against abuse or capricious acts by public officials; a neutral third-party chosen to settle disputes.

oxygen: most common element in the Earth's crust where it combines with other elements; makes up 12% of Earth's atmosphere. It may combine with many other elements to form various oxides.

ozone: a molecule consisting of three oxygen atoms and is a bluish irritating gas with a pungent odor; it is a pollutant in the lower atmosphere and protects the earth from ultraviolet rays in the upper atmosphere. It can be created at ground level by chemical reactions or electric fields.

Pentium: Intel trade name for their 586 microprocessor chip; FAB 11 product.

Pentium chip: a processor chip made by Intel. The fifth-generation of chips that began with the 8086, followed by 286, 386, and 486 types. Because numbers cannot be copyrighted, Intel chose a name in order to prevent competition that occurred with earlier chips.

people of color: a term used to refer to non-white people, used instead of the term 'minority' which implies inferiority and disenfranchisement (communities of color, etc.). The term emphasizes common historical, economic, social and cultural backgrounds.

personal computer (PC): microprocessor based computer system. New models are as powerful as mainframes were 10 years ago.

Petroglyph National Monument: national park adjoining Albuquerque city limits on the West Mesa and located just south of Intel in Rio Rancho; under severe threat of encroachment by development; a sacred site for the indigenous people of New Mexico.

petroglyphs: ancient drawings done on rock by Native Americans.

phosphine: a lethal gas made up from phosphorous and hydrogen. It is used as a dopant in semiconductor wafer production.

plume: refers to the spread of a substance in air or groundwater in three dimensions.

positive photoresist applicators and developers: chemicals used to etch silicon wafers when making chips.

Precursor Organic Compounds (POCs): organic compounds that may become (are precursors to) ozone smog.

protectionism: when producers of a country react to foreign competition in their home market to protect their own products by placing restrictions (as high duties) on foreign competitive goods or subsidies to the local firms. Practiced by all countries-some for selected industries, some for all types of companies.

propane: flammable gas used as fuel or as an aerosol propellant.

Public Service Company of New Mexico (PNM): primary electric utility serving Albuquerque and Rio Rancho.

Pueblo: Pueblo Indian villages in the Southwest.

racism: power plus racial prejudice, a system that leads to oppression or discrimination of specific racial or ethnic groups.

reactive airway disease: disease of the lungs, throat, or nose caused by exposure to toxic substances.

Record of Decision (ROD): a document spelling out a legally binding administrative decision of a government agency, such as remedy selection by U.S. EPA at an NPL site.

red blood cells: part of blood that carries oxygen to other cells in the body.

reproductive toxicity: the characteristic of substances that cause problems with animal or human reproduction such as infertility, miscarriages, birth defects, or problems during pregnancy.

semiconductor: material that is used to carry or conduct electricity, used to make integrated circuits or other electronic components. Silicon is the most common semiconductor.

sewage effluent: the liquid output of sewage treatment plants, as opposed to sludge, the solid output. It is returned to ground or surface waters by sewage treatment plants.

silane: flammable gas made by combining methane and silicon.

silicon: second most common element in the Earth's crust. In nature it is always combined with other elements. Most common combination is silicon dioxide (silica), which forms glass, sand, quartz and other minerals. In pure form it is a semiconductor.

silicon slivers and wafers: pieces of pure silicon used to make integrated circuits. Wafers are usually thin 6 inch circles. Many chips can be made from one wafer.

smokestack chasing: communities trying to attract new businesses to supply local jobs without regard to the negative consequences created by these companies.

sodium hydroxide: strong caustic alkali used to make soap, paper, and to neutralize acids.

solvent handling: equipment and procedures used to handle solvents that may be highly flammable or toxic.

solvent stations: work locations that handle a large variety of solvents.

spiritual interdependence: needs shared by all inhabitants; necessary for a happy and meaningful life.

spreadsheets: computer software used mostly for bookkeeping and accounting purposes. Second most frequent use on PCs after word processing software.

streamlined regulatory permit process: special treatment given to some companies that bypasses or ignores normal health, safety, or environmental safeguards.

sulfuric acid: strong caustic acid used in making explosives, dyes, and fertilizers. Also known as oil of vitriol.

Superfund Amendments and Reauthorization Act (SARA Title III): 1976 amendments to CERCLA that require manufacturers and extractive industries to inform the government about *some* toxic substances that they use. It also requires them to report on substances released into the environment, giving dates, amounts, and other information.

Superfund Law (CERCLA): the Superfund fund only finances cleanup for which the responsible parties cannot be determined or do not have the funds to pay. Most Superfund cleanups are paid for by their owners, operators or other responsible parties (polluters).

Superfund site: common term for sites on National Priorities List, they contain much pollution that affects many people and/or is very expensive to cleanup. When possible, the government attempts to have offending companies pay for cleanup, otherwise, cleanup is done with special taxes and ultimately at taxpayer expense.

sustainable development/sustainable communities: long-term economic development strategy which is developed by communities, results in long-term job creation, protects the air, land and water, and is based in the history and culture of the community (also see economic justice and environmental justice).

systemic intoxication: results of toxic substances that affect the entire body.

tax abatements: lawful subsidies used by companies and individuals that help avoid paying taxes they would normally owe.

tetrafluoromethane: Solvent, flammable, toxic.

thermal inversions: weather condition where air is held in by a larger air mass, becoming stale and contaminated. Common in valleys and during winter months.

thermal oxidation: using heat to break down toxic substances. See In-plant incineration.

toxic emissions: harmful gases or particles released into the air, soil, or water. Toxic emissions are usually considered separately from ozone precursors.

toxic polluters: companies, government agencies or other entities that create pollution and waste that is hazardous to animals and plants.

toxic waste: poisonous by-products of mining or manufacturing.

toxin: anything poisonous to living organisms. Technically a product of the metabolic activities of a life form.

transistors: one of the earlier, simplest form of semiconductors; an electronic valve; an integrated circuit contains the equivalent of two or more (millions) of transistors.

trichloroethane: flammable, toxic solvent, also known as methyl chloroform, it is also a major ozone depleter.

trichloroethylene (TCE): flammable, highly toxic solvent; carcinogenic. Also called "trike."

Utility Expansion Charge: fees charged by public and private utilities to extend lines (water, sewer, electric, phone, etc.) to newly developed areas.

uranium: radioactive element used in nuclear weapons and power plants. Hazardous to plants and animals.

volatile organic compounds (VOCS): substances containing carbon atoms that evaporate very quickly, such as chlorinated solvents.

wafer fabrication: process of making pure silicon that can be turned into integrated circuits; making chips from silicon wafers. See fabrication.

wet chemical stations: work locations designed to handle and use liquids or wet materials.

winter inversions: see thermal inversions.

Workers Compensation Law: requires companies to compensate employees injured on the job.

xylene: flammable, toxic solvent.

Selected Bibliography

The Bureau of Business and Economic Research of the University of New Mexico. <u>New Mexico Tax Study: Phase II Report</u>, 1993.

Bay Area Air Quality Management District. <u>Toxic Air Contaminant Emissions Inventory for 1990-91</u>.

Chavis, Rev. Ben. Speech, First National People of Color Environmental Leadership Summit, Washington D.C., 1991.

Environmental Protection Agency. <u>EPA Toxics Release Inventory, 1987-1992</u>.

Environmental Protection Agency and Regional Water Quality Control Board. <u>Fact Sheets on the Intel Sites</u>, 1991.

Finer, Jeffrey. "Land Subdivision: The Selling of New Mexico." <u>New Mexico People and Energy</u>, 1980.

Hossfeld, Prof. Karen, "Why Aren't High-Tech Workers Organized?" in <u>Common Interests: Women Organizing in Global Electronics</u>. London: Women Working Worldwide, 1991.

"A Labor Relation Review Handout." Rio Rancho, NM: Intel Corporation, 1993.

Ladou, J. "Health issues in the microelectronics Industry: State of the Art Reviews." *Occupational Medicine*, 1986.

Larrabee, Graydon. Speech, International Symposium on Semiconductor Manufacturing, September, 1993.

Lilis, Dr. R. "Diseases Associated with Exposure to Chemical Substances: Organic compounds." <u>Maxcy Rosenau Public Health and Preventive Medicine</u>, 12th Edition, J.M. Last, ed. 1986.

SouthWest Organizing Project. <u>Report on the Interfaith Hearings on Toxic Poisoning in Communities of Color</u>. Albuquerque: SWOP Publications, 1993.

Summers, Lawrence. Memo, December 12, 1991.

Wiegner, Kathleen. "The Empire Strikes Back." *Upside*, June 1992.

Index

New Mexico Occupational Health and
    Safety Bureau 63, 64
New Mexico State Emergency Response Commission   51
New Mexico State Engineer 56, 58, 59
New Mexico Tax Study   40, 46
Noyce, Robert   4, 5

P
Pat Delbridge Associates, Inc.   67
Perlman, Robert   36
Philips/Signetics   20
Public Service Company of New Mexico   37, 45

R
Right to Know laws (Superfund Amendment and
    Reauthorization Act)   51
Rio Grande   2, 31, 40, 59, 64, 65, 66, 75
Rio Rancho   3, 9, 11, 27, 29, 30, 31, 35, 38, 43, 47, 48,
    49, 51, 52, 53, 55, 56, 57, 58, 60, 69, 75
Rio Rancho Utility Corporation   45, 56, 57, 68, 60, 69, 75

S
San Felipe Pueblo   64
Sandia National Laboratories   23, 24, 29, 62
Sandia Pueblo   64, 65
Sandoval County   11, 30, 31, 32, 33, 35, 37, 39, 45,
    46, 47, 64, 73, 77
Santa Ana Pueblo   64
Santa Clara Center for Occupational Safety and Health   21
Santo Domingo Pueblo   64
Semiconductor Industry Association   23
Sheppard, William   57, 58, 68, 69
Shockley Semiconductor   4
Shockley, William   4
Southwest Network for Environmental and
    Economic Justice   37, 151
Sumitomo Sitix Silicon, Inc.   47, 57
Sunwest Bank   33
Superfund Sites (National Priorities List)   12

T
Texas Instruments   9, 50

The Electronics Industry
Good Neighbor Campaign

The Electronics Industry Good Neighbor Campaign (EIGNC) is a collaborative effort between the Southwest Network for Environmental and Economic Justice and the Campaign for Responsible Technology. Participating organizations in the EIGNC include three Southwest Network affiliates, the SouthWest Organizing Project (SWOP) in New Mexico, People Organized in Defense of Earth and her Resources (PODER) in Austin, Texas, and Tonatierra Community Development Institute in Phoenix, Arizona. The other participating group in the EIGNC collaboration is the Silicon Valley Toxics Coalition, based in San José, California.

The EIGNC grew out of the need for a regional, national and ultimately an international community response to the migration of the electronics industry away from the Silicon Valley to low-wage, high public subsidy, low environmental regulation zones in the southwestern United States and abroad. The EIGNC thus brings together local organizations to join efforts in organizing for accountability on the part of the electronics industry towards local communities throughout the region. Each local group identifies its own local programs, and through the EIGNC, works with the others on common or closely related issues. This strengthens the local work of each organization, and makes it possible for the collaboration as a whole to achieve major policy objectives. For example, the EIGNC and its participant organizations played a significant role in obtaining a Department of Defense appropriation for SEMATECH which sets aside $10 million annually for the research and development of worker and environmentally-safe production technologies.

Participant organizations in the EIGNC include:

Campaign for Responsible Technology (CRT)

The Campaign for Responsible Technology was initiated by the Silicon Valley Toxics Coalition in 1991. The CRT brings together labor, occupational health and safety advocates and technology policy advocates with environmental and community activists. The mission of the CRT is to promote democratic grassroots participation in the formation of industrial policy surrounding the development of the electronics industry.

People Organized in Defense of Earth and her Resources (PODER)

People Organized in Defense of Earth and her Resources has been a leading advocate in Austin, Texas communities for protection of communities of color from groundwater and air pollution associated with industrial activities. PODER has been instrumental in 1993 in leading a successful fight to relocate fuel tank farms out of the East Austin area and in monitoring cleanup efforts associated with petrochemical industry storage practices. PODER has also been the leading advocate for safe chemical storage and pollution prevention as the rapidly growing electronics industry expands in Austin. As a part of this work, PODER has focused on employment and community development issues related to SEMATECH, the national public-private consortium of semiconductor companies which includes Intel and which has its headquarters in Austin. PODER is an affiliate of the Southwest Network for Environmental and Economic Justice.

Silicon Valley Toxics Coalition (SVTC)

The Silicon Valley Toxics Coalition arose in response to the severe pollution and workplace safety problems generated by the microelectronics industries in the Silicon Valley of California. The Valley is the birthplace of "high tech" industry and has more documented groundwater contamination than anywhere else in the United States. Groundwater cleanup and protection continues to be the principal work of the SVTC. The organization has been awarded two US EPA Technical Assistance Grants to work on the cleanup of groundwater contamination caused by IBM and the US Navy. In 1992, SVTC launched the Campaign to End the Miscarriage of Justice, an effort which is working to eliminate the use of glycol ethers in the semiconductor industry, and has been active in efforts to eliminate the use of CFCs as well as other chemicals used in semiconductor production.

Southwest Network For Environmental And Economic Justice

The Southwest Network for Environmental and Economic Justice is a regional organization composed of African, Latino, Native and Asian/Pacific Islander Americans and exists to strengthen the work of local organizations and empower communities to impact local, state, regional and national policy on environmental and economic justice issues as these impact people of color. The Network has been developed by representatives of over 75 grassroots organizations from Texas, Oklahoma, New Mexico, Colorado, Arizona, Utah, Nevada and California and Mexico, as well as citizen organizations from Native Nations in the region. The Network, founded in 1990, strengthens poor communities and organizations of color under stress from environmental degradation and economic injustice in the Southwest and in Mexico. The Network includes community, youth, student, human rights and labor organizations, and works closely with many organizations throughout the United States working to address issues of environmental and economic justice.

SouthWest Organizing Project

SWOP, founded in 1981, is a multi-racial, multi-issue statewide membership organization in New Mexico whose mission is to empower the disenfranchised in the Southwest to realize racial and gender equality, and social and economic justice. SWOP's work focuses on increasing citizen participation in communities of people of color in the basic decision making which affects our lives. Current SWOP direct organizing efforts in New Mexico are working to promote accountability among industries, the military, and governments at all levels on a range of environmental and economic justice issues, as well as promote greater community control over industrial, commercial and residential development. This work takes place at both the local and state levels, and includes activities ranging from neighborhood organizing work to statewide campaigns. SWOP is also a founder and affiliate member of the Southwest Network for Environmental and Economic Justice.

Tonatierra Community Development Institute

Tonatierra is a community-based organization in Phoenix, Arizona which grew out of the farmworker justice movement and which focuses on a range of development, educational and cultural issues of importance to indigenous communities. Tonatierra has been building coalitions with community-based and environmental groups concerned about the cleanup and protection of aquifers polluted by Motorola Corporation, which is presently the largest employer in the state. Tonatierra has produced a video entitled *Silent Waters* which the organization is using for educational purposes, and most recently participated in air permit hearings and other activities related to the siting in April of Intel's new P6 chip FAB. Tonatierra is a founder and affiliate member of the Southwest Network for Environmental and Economic Justice.

Also Published by SWOP:

500 Years of Chicano History In Pictures
500 Años del Pueblo Chicano
Paperback: $16 • Hardback $35
(plus postage: $2 paperback; $2.50 hardback; NM residents only, add tax: $.93 paper; $2.05 hard)

This very important bilingual book of Chicano history contains over 800 photos, the text is in Spanish and English. This is a wonderful book for people of all ages. It's a great way for Chicano youth to learn and understand our history and culture. It's also important for people of all ethnicities in order to better understand Chicano history. This book is published by the SouthWest Organizing Project and edited by Elizabeth Martínez, noted author of books and articles on Latino issues.

NOW AVAILABLE:
Curriculum Guide for "500 Years of Chicano History"

SWOP has produced a very special Educational Curriculum Package which includes: A "Curriculum Guide for Elementary and Secondary School Teachers" (authored by Judy Drummond and Susan Katz Weinberg); a paperback copy of the book; AND the 2-part Video "Viva la Causa! 500 Years of Chicano History." The 70 page Guide is divided into 12 themes and presents a synopsis and a sample lesson plan that will motivate students to participate in discussions and activities aimed at learning the Chicano experience. Also provided are other interesting ideas and resources for your classes.
Curriculum Package—ONLY $112⁵⁰

ANNOUNCING A BRAND NEW VIDEO TAPE:
¡Viva La Causa! 500 Years of Chicano History

A two-part educational documentary video based on the book 500 Years of Chicano History in Pictures/500 Años del Pueblo Chicano by Elizabeth Martínez. Both parts are 28 1/2 minutes each, they depict pre-Colombian times to the present. The video incorporates photos from the book, archival footage, narrators, and lively music from corridos to rap. It is directed by Elizabeth Martínez and Doug Norberg, and produced by Doug Norberg. It is a joint project of the SouthWest Organizing Project and Collision Course Video Productions of San Francisco, California. $37⁵⁰ (free postage).

To Order: Send check or money order with your name, address, telephone number and describe the item you are purchasing. Send to:
SouthWest Organizing Project
211 10th St. S.W., Albuquerque, NM 87102 • (505) 247-8832, FAX (505) 247-9972